P9-CEM-044

DO NOT STEAL
THIS BOOK
BOB CONQUEST

Behavior Analysis and Systems Analysis

Contributors

John M. Atthowe, Jr.	*Professor of Psychology Rutgers University Medical School*
Frank Baker	*Assistant Clinical Professor of Psychology Harvard University Medical School*
Anthony Broskowski	*Senior Planner, United Community Planning Corporation, Boston, Massachusetts*
James Ciarlo	*Director, Mental Health Systems Evaluation Project, Denver General Hospital Mental Health Clinic*
Harold W. Demone, Jr.	*Executive Vice-President, United Community Planning Corporation, Boston, Massachusetts*
Donald H. Ford	*Dean, College of Human Development, Pennsylvania State University*
Dwight Harshbarger	*Associate Professor of Psychology and Acting Director of Clinical Training, West Virginia University*
Jack A. Horrigan	*Research Associate, Denver General Hospital Mental Health Clinic*
Alan Kazdin	*Assistant Professor of Psychology, Pennsylvania State University*
Carolyn H. Kemp	*Chief, Koko Head Clinic, Honolulu, Hawaii*
Donald C. Klein	*Program Director, Center for Macrosystem Change, NTL Institute for Applied Behavioral Science*
Jon E. Krapfl	*Associate Professor of Psychology and Director, Professional Masters Degree Program, West Virginia University*
David Levine	*Chairman and Professor of Psychology, University of Nebraska*
Leo Levy	*Associate Professor, Department of Preventive Medicine, University of Illinois Medical Center*
Roger F. Maley	*Chairman and Associate Professor of Psychology, West Virginia University*
Richard W. Malott	*Associate Professor of Psychology, Western Michigan University*
Terence P. Smith	*Manpower Development and Training Specialist, Department of HEW, Boston, Massachusetts*
Fred E. Spaner	*Chief, Community Mental Health Consultation Section, National Institute of Mental Health*
Leonard P. Ullmann	*Professor of Psychology, University of Hawaii*

Behavior Analysis and Systems Analysis: An Integrative Approach to Mental Health Programs

Edited by:
DWIGHT HARSHBARGER
ROGER F. MALEY
West Virginia University

Behaviordelia, Inc.
Kalamazoo, Michigan

Printer: Edwards Brothers, Inc.
2500 South State Street
Ann Arbor, Michigan 48104

Standard Book No.: 0-914-47411-1.

Contents

Preface

Under the auspices of the Department of Psychology at West Virginia University, the subject matter of community-clinical psychology has been the focus of West Virginia Conferences in 1970, 1971, and 1973. The first two conferences focused on the psychology of private events and the use of the group as an agent of change. The current volume, based on the 1973 Conference "Issues and New Developments in the Delivery of Mental Health Services", addresses the problems of behavior analysis and systems analysis and suggests an integration of these areas. This Conference's theme recognizes the need for viewing the increasingly sophisticated technologies of behavior change within a systems perspective. Our clinical procedures can no longer be viewed as somehow standing apart from the overall organization of services and dynamics of delivery systems.

This volume reflects the efforts of some of the most productive and vigorous professionals in both systems analysis and behavior analysis. It is a testimony to the vitality of both areas and a statement of faith that the combination of the two will take us several steps closer to designing and executing more effective human service delivery systems.

It is our hope that this volume will be more than a vehicle for disseminating knowledge and viewpoints. We trust that it will provide an impetus for the generation of research and services which take into account the strengths of the two conceptual systems. At present we simply do not know how powerful and wide-spread this integration might become. Mental health professionals will judge the success of our efforts by their actions in implementing the ideas and procedures reflected in this volume. Whether successful or not, we eagerly anticipate your vigorous reactions.

Acknowledgements

The editors are grateful to our sponsoring institution, West Virginia University. We are especially indebted to Dr. Ray Koppelman, Provost for Research and Graduate Affairs, who so ably helped us in the preparation of the Conference and even seemed to enjoy himself while attending the Conference and welcoming the participants. The assistance of many others associated with West Virginia University, particularly the faculty of the doctoral program in clinical psychology, was greatly appreciated. The Conference Office at West Virginia University was instrumental in making the Conference an enjoyable three days.

The secretarial skills of Mary Ann Watson and Bonnie I. White were essential in both the organization of the Conference and the preparation of the manuscripts.

And finally, to the Conference participants and the contributors to this volume we acknowledge a very special thanks. We thank them not only for their scholarly products, but also foı their cooperation, warmth, sense of humor, and belief in the importance of this enterprise. They made our work enjoyable and edifying. We hold them in high esteem.

SECTION I

The Dilemmas of Deviance and Intervention

A very few years ago it would have been considered naive to question if the mental health professional should *be engaged in changing people's behavior and attitudes. Yet, today, many critics, who are questioning this most basic of all assumptions, have emerged. The recurring issue of control and counter-control, the dilemma posed by the possible psychological coercion of political dissenters, the arguments as to the best locus of change-efforts, and the ethical issues surrounding planned social action are recurring problems which increasingly force themselves into the consciousness of psychologists.*

If mental health professionals veiw the community as the client, and work to change the community in ways which reduce environmental stress, they have committed themselves to a strategy of changing environments rather than people. The engineering concerns surrounding a community focus for mental health activities revolve around our knowledge of how communities really work and how to go about changing them. Do we have a viable theory of community intervention, let alone reliable empirical data? And, perhaps more basically, do we have evidence that social change does away with emotional and behavioral disorders?

To put it another way, what is the precise relationship between community factors and the development of psychopathology?

In Levine's chapter, "The Dangers of Social Action", the focus is on planned social action and the dangers involved in mental health professionals becoming actively involved in this arena. It seems clear that the last decade has witnessed the movement of the traditional mental health professions from helping individual patients adjust to society, to social-system innovation, and change strategies. Even though we know that environmental influences produce, or are significant factors in, most problems of living, do we know enough to systematically go about changing society? Do we have the "right" or the "mandate" to change society? Might we be co-opted into making either undesirable changes; or worse, meaningless changes, which are undertaken to provide illusions of dramatic change and, consequently, reduce the pressure for change? Obviously, these are extremely complex ethical and engineering problems which transcend the training and expertise of most of today's mental health workers.

One can only conclude, with Levine, that our intervention mistakes at the societal level will hurt many more people than our blunders in individual therapy. It still is not clear that community psychologists should be given the types of societal responsibilities and powers that have traditionally been reserved for political and legal processes. However, the dye has probably been cast and we will, for better or worse, continue to muddle through these murky waters with or without divine guidance.

The Harshbarger and Maley chapter, "Getting off the Horns of a Dilemma: Behavioral Methods of Treatment", provides some relief from these problems by suggesting that the behavioral model of treatment has several features which generate significant improvement over the psychodynamic treatment model. In general, the behavioral position is more likely to generate public verification of the treatment process, to provide more outcome-oriented delivery systems, and to have a firmer commitment to accountability and the design of rational feedback systems than psychodynamic models.

Behaviorally based treatment systems reduce the likelihood that therapists inadvertently act as political agents for the social control of deviance. The value system of the client and his specification of the problem are more likely to be accepted. Social class biases are diminished when treatment plans truly incorporate changes defined as meaningful by the client. Merely by collecting information on, and attending to, specifiable target behaviors, the behavior therapist can more rationally decide on the process and the termination of therapy. The overall quality of delivery systems is improved by the possibility of product outcome analysis which may lead to significant changes. It is only in this way that treatment technologies can be evaluated in terms of their relative cost-effectiveness. Thus, it seems very likely that in the future, should mental health professionals be required to become more accountable for their activities, the behavioral model will be far more helpful than psychodynamic models of treatment.

If one looks carefully at the huge costs

of our society's effort to improve the mental health of its citizens, and if one considers the questionable or undocumented effectiveness of most programs, the demand for reassessment and reorganization is clear. Ford's chapter, "Mental Health and Human Development: An Analysis of a Dilemma", stresses the fragmentation of our mental health efforts and proposes that we give up on the use of the concept of "mental health", because we simply cannot adequately define its domain. "Mental health" is a discipline and guild related term, according to Ford, who advocates a new framework for essential integration across disciplines and professions.

Developmental and preventive approaches to human services should take their place alongside remedial approaches. Effective correlation of multiple approaches to human service delivery systems is a pressing need. The newly emerging patterns for the delivery of services must erase the barriers which currently exist among separate, categorically defined, and guild controlled service systems. We, as professionals, must stop the defensive posturing associated with protecting our turfs and identities and begin to work toward these new, more effective arrangements.

David Levine

Most mental health professionals believe that they should be actively engaged in the arena of planned social and political action. In the following chapter, Levine points out the dangers inherent in the establishment of a mental health power elite which is not accountable to the general public and not aware of the risks involved in "community psychology". Thus, Levine joins a respectable list of psychologists who caution us to beware of encouraging suffering people to adjust to an evil society; to avoid deceiving ourselves concerning our lack of expertise in changing organizations; and to be careful of being used by other people who wish to merely control others or to maintain the status quo. According to Levine, our new techniques and commitments to change the environment, not the individual, have put community practitioners on a very shaky and unfamiliar ground. Have any of the mental health professions really thought through the implications and dangers inherent in their attempts to change large segments of society?

To minimize the risk of disastrous intervention efforts, Levine believes that community psychologists need broad-based interdisciplinary training in history, sociology, law, economics and political science, as well as humility in regard to psychology's role in solving the world's problems. Community psychologists will undoubtedly rush into the fray of social change, and warnings such as this might help us to behave more responsibly and rationally in this most sensitive and complicated arena.

CHAPTER 1

The Dangers of Social Action

The mental health frontier is shifting from the amelioration of illness to preventive intervention at the community level. The community itself is being taught to collaborate in creating health-giving environments . . . Reference was made to optimal realization of human potential through planned social action.

Bennett, Chester C. "Community Psychology: Impressions of the Boston Conference on the Education of Psychologists for Community Mental Health." **American Psychologist,** *1965.*

Obsessive reformist delusions are a symptom of mental illness requiring hospitalization.

Snezhnevsky, Andrei V., Chief Psychiatrist of the USSR Ministry of Health and Secretary of the Academy of Medical Sciences at a meeting of the Ministry of Health, June 12, 1970. (As reported by Andrei Dmitriyevish Sakharov in Medvedev, A. & Medvedev, R. **A Question of Madness.** *New York: Knopf, 1971.)*

The purpose of this paper is to explore the possibility that Bennett's views might be the first step in the development of a theoretical position which could culminate in the kind of statement attributed to Snezhnevsky — the possibility that by entering the arena of planned social action, mental health professionals are in danger of establishing themselves as a power elite, functioning autonomously and with little opportunity for public scrutiny and public accountability.

Thinking through this possibility will have implications for whether or not psychologists should involve themselves in attempts to organize community resources for the delivery of mental health services. If the analysis leads some readers to believe that psychologists should continue in this involvement, it may help them involve themselves in ways which are more effective, efficient, productive and — perhaps most important — ethical and democratic.

HISTORICAL BACKGROUND OF COMMUNITY PSYCHOLOGY

The historical background of *clinical psychology* is well known (Reisman, 1966; Watson, 1953), and there is little in it which reflects a community perspective until the mid-fifties. Before that time, clinical psychologists saw themselves as filling three primary functions: psychological testing, psychotherapy, and research. If anyone was doing *community psychology*, it was those social psychologists who developed and promoted the Society for the Psychological Study of Social Issues. Beginning with the mid-fifties, however, clinical psychologists became increasingly dissatisfied: some weren't sure that psychotherapy worked; psychological testing seemed useless and boring — nobody read the reports anyway; research was considered irrelevant to the solution of clinical problems. At the University of Nebraska, we evolved an approach we called *community clinical psychology* which seemed to help us overcome these dissatisfactions — or perhaps just made it easier for us to live with them.

As described in Jones and Levine (1963) and Levine (1968a), we found that by emphasizing the social context we were able to gain a more complete understanding of the problems with which we and our clients were trying to cope: we were able to see more readily why psychotherapy worked at one time and not at another; why our test reports might not have been as useful as we hoped; and what kind of research and scholarship might be relevant to solving clinical problems. Our approach, however, remained focused on the individual; we simply kept reminding ourselves that the individual could only be understood as functioning within various, often complex, social contexts. Of course, we are still struggling to define precisely what we mean by the term "the broadest social context" (Levine, 1968a).

At about the same time, however, other psychologists were responding to their dissatisfaction by taking more radical steps — steps which culminated in the Boston Conference (Bennett, 1965) and the emergence of *community psychology*. Community psychology removes the individual from the center of attention. The goal of community psychology is to improve human welfare, but the technique is to change the environment, not the individual.

For the purposes of this chapter it is useful to be clear about the distinction I am making between community psychology and community clinical psychology. As I see it, the distinction stems primarily from the extent to which attention is paid to the *individual*. In community psychology there is a commitment to community organization, social change, working with decision-makers, social planners, and politicians. This is based on the notion that only by changing society, the community, or large organizations, can one expect to cope with the massive problems of crime, "mental illness", and other *personal* difficulties.

Community clinical psychology, on the other hand, maintains the individual as the focus of attention. Its contribution over traditional clinical psychology is the emphasis on the need to recognize the social context (ecology) in which the person lives and works.

Other writers have made a distinction similar to the one that I am proposing. Cowen (1973), for example, uses the term "community mental health" in the way that I am using the term community clinical psychology: "This point bears on the oft-cited difference between 'community mental health' and 'community psychology'. The former tends to accept (diagnostic) entities as givens and to use diversified community services and approaches to augment effectiveness in coping with them. The latter, however, is broader and more oriented to engineering health-producing settings and environments."

Some writers, like Sarbin (1970) and Lipton and Klein (1970), prefer to use the term community psychology to include the total range of intervention strategies from changing individuals through changing organizations to changing society as a whole. At one point, Sarbin writes, "The targets (of community psychology) are *organizations* of persons whose conduct is dysfunctional as a result of ecological misplacement or nonplacement." This seems to place him toward the community psychology end of the scale until we look further. On the next page he writes, ". . . the focus is on the individual as a *member* of a collectivity, acting and interacting, with others." Now he seems to be more at the community clinical end.

Lipton and Klein's (1970) definition is especially broad: " 'community psychology' therefore is concerned with the reciprocal relation of individuals and their environments, and with the development of ways in which interventions designed to foster more adequate person-environment relationships can best be effected."

However, there are psychologists who emphasize "social change" as the major contribution of the community psychologist: "Although the community psychologist can work effectively at a variety of levels, his most significant endeavors will be to bring about social system innovation and constructive change." (Cook, 1970.)

DANGERS OF SOCIAL ACTION

We can, therefore, conceptualize intervention as taking place at three possible loci: the individual, organizations of individuals within society, and society as a whole. Individual psychotherapy, desensitization therapy, and behavior modification are illustrations of intervention at the individual locus. At this point, the psychologist is likely to be talking about focus on the individual child (rather than adult), or about crisis intervention rather than long-term therapy. The locus, nevertheless, is on changing the individual so that he can better cope with his environment (ecology).

At the organizational locus, we have the community psychologist serving as a consultant and helping organizations change themselves to alleviate individual stress without changing the organization's goals. This is the locus at which most community psychologists now function: working toward structural changes in school systems, mental and correctional institutions (like token economies), increasing community control of comprehensive mental health centers and other human service agencies, and so on.

I know of no psychologist, qua psychologist, working to change society as a whole. It is possible, of course, that this notion of "changing society" is mythical — that is, that it is possible on an *abstract* level to conceive of an entity we label "society" and since *entities* can change, therefore expect that "societies" can change. But it is possible that "society" is a functioning amalgamation of separate structures, each of which is made up of large numbers of individuals and, as such, does not change as a whole, but rather evolves through changes in its subsystems.

If this is so, then our conceptualization might be more heuristic if we conceived of a continuum from, at one end of the spectrum, changing organizations which have slight impact on few numbers of people to, at the other end, changing institutions which have major impact on large numbers of people.

Dangers at the Locus of the Individual

Although the theme of this chapter concerns the danger of what I am calling community psychology, it is important to note the drawbacks of community clinical psychology as well. The most troubling danger at the locus of the individual stems from the historical problem of defining adjustment: community clinical psychology may be open to the criticism that it encourages the suffering person to adjust to an evil society rather than work to change it.

According to Hurvitz (1973): "Psychotherapists have created a community mental health movement and have established programs, facilities, and activities designed to quiet restive neighborhoods." ". . . psychotherapy creates powerful support for the established order — it challenges, labels, manipulates, rejects or co-opts those who attempt to change society."

Simon's (1970) criticism is broader in scope: ". . . since its inception, and despite the explicit liberalism of many of its practitioners, clinical psychology has been committed to a political perspective that has consistently led it away from a productive involvement in social change."

Those psychologists whose goals and techniques lead to the pacification of active attempts to change the social order may be helping the individual (or not, as the case may be) — but Hurvitz and Simon argue convincingly that they are not improving human welfare. Moving beyond community clinical psychology to community psychology, they urge social change. This step, however, entails risks, and it is these risks to which I will now turn.

Dangers at the Locus of Changing Organizations

The first possible danger psychologists may create when they focus on changing organizations rather than changing individuals is that they may not know what they are doing. Bennis (1970) suggests that there is no viable theory of social change. To the extent that this is true, our suggestions to organizations are based not on the application of scientific knowledge, but rather on some mix of a particular psychologist's values, bias, wisdom, folklore, and experience. Psychologists who lack wisdom, restraint, and good judgement may be dysfunctional and harmful.

A second danger arises when, focusing on changing organizations, psychologists deceive themselves that they are improving society and human welfare when, in fact, they are, as Hurvitz suggests, strengthening the status quo by ameliorative measures which take the steam out of a discontented group. Helping an organization "improve the morale" of its workers, students, or inmates does not change the distribution of money or power; the "troublemakers" are co-opted or suppressed. The sticky issue for the psychologist is deciding when an organizational change is a significant social improvement and when it is a meaningless compromise — or worse.

Related to this second danger is the entire issue of the relationship between the generally white middle-class professional community psychologist and the community people whose lives he is so benevolently trying to improve. Much has been written about the need to understand and respect the differences in the cultural background of these two groups. It is also important to remember the different self-interests of these two groups. Meaningful social and organizational change and complete community control may put the community psychologist out of work.

The interpersonal relationship between professionals and community representatives can become a profoundly complex situation.

Hersch (1972) describes some of the problems he has encountered with community control:

> For one thing, community control situations often bring forth the emergence of local leaders who are not, as they are expected to be, representative of the community . . . In addition, some of the local leaders turn out to be among the more psychopathological members of the neighborhood . . . Another unforeseen problem is that power on the community control board is often not wielded for the sake of the rational development of the program. Frequently, self-interest may take priority over program needs . . . Another unexpected feature is . . . the intense anger that often emerges . . . directed at the professionals themselves . . . Finally, one often has the feeling that power on the board is being exercised for its own sake . . ."

Hersch's paper is too recent to have received much response as yet. When that response comes it is likely to be intense: as if the community psychologist never wields power "for his own sake", or never out of "self-interest" or never from "intense anger".

Saul Alinsky's (1965) view of the relations between the professional and the poor is, predictably, diametrically opposed to Hersch's:

> This (effective social change) means an organized poor possessed of sufficient power to threaten the status quo with disturbing alternatives so that it would induce the status quo to come through with a genuine decent meaningful poverty program. After all, change usually comes about because of threat, because if you don't change, something worse is going to happen. Rarely in history do we find that the right things are done for the right reasons such as an anti-poverty program launched on a moral dynamism. It is always done for other political reasons.

The final danger at this locus of intervention is that the errors we make will affect many people. Bad individual psychotherapy hurts one person; a bad token economy hurts an entire ward of people; bad social change hurts a countless multitude. Although we can make outright mistakes in recommendations for organizational change, the more common failing on the part of community psychologists may be the making of recommendations on the basis of incomplete — rather than inaccurate — information. This is a particularly thorny problem since "no decision" is one kind of decision, and one may argue that a decision based on minimal data is better than a decision based on chance alone.

The danger here may come from two directions. First, the input from the social scientist may be given excessive weight in comparison with other input by the responsible government official (or other decision-maker). The decision-maker usually has no way of knowing whether the grounds for the psychologist's recommendations are flimsy or firm. This can be especially troublesome for those psychologists who appear as expert witnesses. It is sometimes difficult to communicate to an attorney or judge the degree of confidence in an expert opinion.

The second problem is more directly attributable to the psychologist: he may overvalue his own contribution to the decision in comparison to other contributions. As the clinical psychologist emerges from under medical domination, he runs the risk of making the same kind of mistake some physicians have made. Psychologists had much to contribute to "medical" decisions, but were often ignored. Now psychologists need to realize that politicians, philosophers, lawyers, parents, taxpayers also have wisdom concerning the means of solving social problems.

Dangers at the Locus of Major Social Change

Once the psychologist moves in the direction of changing major organizations or of changing "society", two dangers become apparent. The first deals with the possibility that he will be used by others to achieve evil goals, the second, with the possible dangers of his being put in a position of power.

Since Hiroshima no scientist can easily take the position that he is concerned only with pure science — that the application of the knowledge he develops is not his concern. Once having taken this first step toward involvement, however, the psychologist cannot avoid facing the fundamental questions of human life. The controversy between Rogers and Skinner, between the Humanists and the Aversive Conditioners, between Ellis and Sarbin, are not scientific in nature. When followed through analytically, they emerge as ethical dilemmas.

Perhaps Anthony Burgess in *A Clockwork Orange* posed the problem as well as anyone: If (and, of course, this particular question is still a hypothetical one) a psychologist could change a ruthless murderer and rapist into a passive, meek person who could not defend himself — a person who had lost the capacity to "choose" to do evil — should the psychologist make this knowledge available? Should he make the knowledge generally available? Available only to his government? Only to an industrial firm?

Although one may argue that this question will never in actuality confront a psychologist, there are many similar kinds of dilemmas which do: Does a psychologist participate for the prosecuting attorney in a homocide case even though he disagrees with the nature of the punishment which our judicial system invokes? What about the psychologist who believes — on the basis of his interpretation of available data — that Blacks do not learn best in the same system as Whites, but also believes that segregation of schools will have deleterious effects on the personality of black children?

Richard Christie (1973) assisted the Scranton Seven defense to select a jury which would be more likely to acquit the defendants. Would he have as readily assisted the prosecution? On the basis of the contradictory literature on moderator variables, does the psychologist recommend different cutting scores for Blacks and Whites on academic selection tests? What about the use of personality tests for personnel selection — especially when used by government? Doesn't the necessity for answering personal questions, in fact, constitute a dangerous invasion of privacy?

There seems hardly an instance of social involvement which does not bring with it a complex ethical dilemma for the community psychologist — and yet psychologists rarely get additional training in the analysis of legal or ethical problem situations. Elsewhere (Levine, in press) I have discussed, at length, the fact that people seeking power or trying to maintain their positions of power have used psychologists or psychiatrists to help achieve their ends. Among the illustrations of this tactic was Defense Secretary McNamara's statement "that a psychiatric examination was given to a Navy officer who discussed the 1964 Gulf of Tonkin incidents with Senator J. W. Fulbright." Although Mr. McNamara said the examination "can in no sense be viewed as an act of intimidation or reprisal" against the officer, he also said that disclosure of the report's details "would undoubtedly be harmful to the officer." (*International Herald Tribune*, Feb. 28, 1968.)

The final danger — and the one toward which I have been heading — is the possibility that community psychologists may seek or be offered positions of special responsibility simply because they have the credentials of "psychologist"; that somehow or other, society will be led to believe that as a function of their special training, experience, or knowledge, psychologists should be given positions of greater responsibility and authority. Community psychologists want to change society and society may be willing to give them special authority to implement this change.

But let's try to imagine what happens next. What are some of the social changes community psychologists are talking about? When they enter the political arena what are their concerns? How will they achieve these ends? What do they want to change society to?

Recent events lead me to believe that psychologists — like most people — are most politically active when their own special interests are at stake. The profession has worked especially hard to achieve coverage by medical insurance contracts; to influence legislation so that psychologists may be permitted to serve as Directors of Comprehensive Community Mental Health Centers; to make certain that payment to psychologists qualifies as an income tax deduction. Although there is nothing evil in promoting self-interest, these are certainly not the grand social changes we have been talking about.

Clearly, community psychology is hoping to bring about changes which will increase the delivery of human services, facilitate equal protection by the law, insure equal opportunity in employment and education, and "promote human welfare". But even the new graduate student in Political Science — not to speak of the ordinary citizen when he thinks about it — understands that these kinds of changes are difficult to achieve. They are difficult to achieve because a sizable number of people do not want these changes to occur. A powerful segment of society is satisfied with the way things are now. The experienced politician tells us the "people vote their pocketbook".

Does the community psychologist have any expertise in overcoming this kind of opposition? Bennis says, "no," and John Gardner, in becoming Chairman of Common Cause has, by implication, said that it is not through research and professional practice that one changes society. It is through political lobbying and legal action; I don't see graduate training in psychology as relevant preparation for this kind of activity.

But, one may argue, it does no harm to have a Ph.D. in Psychology as background for this job; scholars have special training and can make an important contribution. On the contrary, it seems to me important to remember that scholars tend to become specialists and that a specialist in one area has no claim to expertise in another area. Further, science is a never-ending process; data and explanations which seem correct today may prove deficient tomorrow. Even when it seems absolutely clear that a scientist knows more than anyone else about a specific problem, considerable care should be taken not to allow him to assume too large a share of the responsibility for social planning and decision-making. When a person, or group of people, accepts an inordinate share of social responsibility, he, or they, may lay claim to special powers or privileges and, by scarcely perceptible stages, alter a democratic form of government. B. F. Skinner in *Walden Two* and George Orwell in *1984* present contrasting fictional illustrations of, and attitudes toward, such a development (Levine, 1968b).

CONCLUSIONS AND RECOMMENDATIONS

I have proposed a distinction between community psychology and community clinical psychology, and a continuum from intervention focused on changing individuals, through intervention focused on changing organizations, to intervention aimed at changing society. I have suggested that the important step toward social action which characterizes community psychology carries with it — along with tremendous potential benefit — some serious risks.

These risks include:

1. At the locus of the individual — that it encourages the suffering

person to adjust to an evil society rather than work to change it;

2. At the locus of organizations — that we do not know enough to be useful; that we may be deceiving ourselves that the change is a major social innovation when, in fact, it is a meaningless compromise; that interpersonal relations between the professional and the community people are especially complex; and that any errors we make hurt more people;
3. At the locus of major social change — that the community psychologist may be used by others to achieve evil goals and that he himself may be put in a position of social and political power solely by virtue of his professional position.

To minimize these risks, I recommend:

1. That graduate training programs in community psychology become interdisciplinary in nature — there is need for background in history, sociology, law, economics, and political science.
2. That simultaneously, these programs strengthen their offerings in statistics, test-design, research methodology, and program evaluation.

   That, as a result of the above recommendations, it will probably be necessary to increase the time needed for the Ph.D. in community psychology — or better yet to conceive of community psychology in terms of a Post-Doctoral Training Program.

My final recommendation is a plea for humility. Psychology can't solve the world's problems. If society is sick, it might be wise to follow Shakespeare's advice and remember that "therein the patient must minister to himself." In any event, social action in a democracy needs the approval of a majority of the people.

Although I have written about the dangers of community psychology, I hope I will not be misunderstood. I believe that there is much potential in the notion of a social scientist contributing to constructive social change, both as scientist and as private citizen. But the risks of this kind of involvement are there and should not be minimized.

Perhaps, however, the dangers are not as great as I fear. A sentence from the 1971 report of the Interdepartmental Committee for Atmospheric Sciences may be paraphrased to read: "What the public thinks about the modification of behavior and the modification of society, rather than what the social scientist knows about it, will play the dominant role in the future of this science."

REFERENCES

Alinsky, S. D. "The war on poverty – political pornography." *Journal of Social Issues*, 1965, 21, 41-47.

Bennett, C. C. "Community psychology: Impressions of the Boston Conference on the education of psychologists for community mental health." *American Psychologist*, 1965, 20, 832-835.

Bennis, W. G. "Theory and method in applying behavioral science to planned organizational change." In P. E. Cook (Ed.), *Community Psychology and Community Mental Health.* San Francisco: Holden-Day, 1970.

Christie, R. *Division Eight Newsletter*, 1973.

Cook, P. E. *Community Psychology and Community Mental Health.* San Francisco: Holden-Day, 1970.

Cowen, E. L. "Social and community intervention." *Annual Review of Psychology.* 1973, 24, 432-472.

Hersch, C. "Social history, mental health, and community control." *American Psychologist*, 1972, 27, 749-754.

Hurvitz, N. "Psychotherapy as a means of social control." *Journal of Consulting and Clinical Psychology*, 1973, 42, 232-239.

Interdepartmental Committee for Atmospheric Sciences, 1971 Report, *Science*, 1973, 180, 1347.

Jones, M. R. & Levine, D. "Graduate training for community clinical psychology." *American Psychologist*, 1963, 18, 219-223.

Levine, D. "Why and when to test? The social context of psychological testing." In A. I. Rabin (Ed.), *Projective Techniques in Personality Assessment: A Modern Introduction.* New York: Springer, 1968 (a).

Levine, D. "Social responsibility of social scientists." *Current Anthropology*, 1968, 9, 417 (b).

Levine, D. "Crime, mental illness, and political dissent." In June L. Tapp (Ed.), *Socialization and the Law.* (in press)

Lipton, H., & Klein, D. C. "Training psychologists for practice and research in problems of change in the community." In Daniel Adelson and Betty L. Kalis (Eds.), *Community Psychology and Mental Health: Perspectives and Challenges.* Scranton, Pa.: Chandler, 1970.

Medvedev, Z., & Medvedev, R. *A Question of Madness.* New York: Knopf, 1971.

Reisman, J. M. *The Development of Clinical Psychology.* New York: Meredith, 1966.

Sarbin, T. R. "A role-theory perspective for community psychology: the structure of social identify." In Daniel Adelson and Betty L. Kalis (Eds.), *Community Psychology and Mental Health: Perspectives and Challenges.* Scranton, Pa.: Chandler, 1970.

Simon, L. J. "The political unconscious of psychology: Clinical psychology and social change." *Professional Psychology*, 1970, 331-340.

Watson, R. I. "A brief history of clinical psychology." *Psychological Bulletin*, 1953, 50, 321-346.

Concepts of deviance and therapeutic strategies aimed at "treating" deviance in mental health systems have been strongly rooted in the prevailing values of communities. As members of their communities, therapists, it has been argued, act as agents of social control when they engage in what they believe to be helpful therapeutic endeavors. This problem is further complicated by the fact that psychodynamic therapies tend to be heavily value-laden, further intensifying the interference of social values in the therapeutic process. Increasingly, mental health professionals are asking if, in their attempts to be helpful, they are doing harm to their clients.

In the following paper Harshbarger and Maley attempt to unravel some of the complex threads of this dilemma. They suggest that if one designs treatment strategies within the confines of traditional therapeutic approaches the dilemma is likely to continue, a situation which is of little service to therapists, clients, or communities. However, they suggest that there are some ways of resolving these problems, and specify how behaviorally based methods of treatment are likely to move mental health treatment systems off the horns of this dilemma.

Dwight Harshbarger

Roger F. Maley

CHAPTER 2

Getting off the Horns of the Dilemma: Behaviorally Based Treatment Systems

The notion of human intervention raises a host of problems, not the least of which is that, as a wizened old mountaineer put it, "people don't want to be intervened on." The fact remains that they are, and that much of what is called human intervention is related to what is called mental health.

THE DILEMMA

It is the dilemma of those who would do good in this world that their efforts must also be viewed in terms of their potential harm. Charges have been levied against the practitioners and managers of those systems of treatment, placed under the rubric of mental health, that they are unwitting agents of political and social control for the larger society. As such, it is alleged, they act to preserve and protect those behaviors that are sanctioned within the tolerance limits of middle class society and its value systems.

The treatment dilemma stemming from the issues of social and political control have been presented by such writers as the Braginskys (1973), Kenniston (1968), Levine (1964), Scheff (1966), Szasz (1961) and many others. The essential elements underlying the dilemma, as developed by this group, are perhaps best put by the Braginskys. Following a review of recent research on the influence of the personal values of therapists and clients on treatment decisions they state:

> We could go on enumerating the foibles of psychologists and psychiatrists forever. But that would miss the basic point. The basic point

> is that the entire system within which psychologists and psychiatrists operate is socio-political. The entire system of psychological classification, and labeling, and diagnosis is faulty.

They go on to suggest that in surmounting the problems to human contact it would be helpful, "If psychologists and psychiatrists were humble, if they met their clients as equals, if they faced the human condition without trying to classify it by the prevailing ideology, then perhaps they could conceivably surmount this barrier."

Private Models of Treatment

Just as we have historically assumed that the locus of deviance rested in an individual's private mental events, we have also legitimized the locus of treatment as resting in the private and professionally protected world of treatment practitioners' mental events. Whether operating in the closed system depicted in *A Clockwork Orange*, or the private client-practitioner relationships of the psychotherapist, treatment systems have been closed relationships, owned and operated by practitioners.

It is not as if mental health treatment practitioners have politically maneuvered to create and gain the advantage in this closed system. Rather, they generally have known no other. Their theories, the organizations in which they received their training, and the service delivery systems in which they work, reflect a set of theoretical constructs based upon the values of the middle class. Generations of psychotherapists have been trained in a theoretical system having its roots in the lives of middle and upper-middle classes in nineteenth-century Vienna. More recently this has been supplanted by the existential problems of middle-class America (R. May, 1969; 1972).

Treatment models such as these are likely to continue for some time to come. Despite evidence that long-term psychotherapy is a relatively ineffective treatment strategy, it continues to be widely used in both practitioner training programs and mental health delivery systems. Stevenson noted in 1959 that personal conviction and tradition, rather than data-based evidence, were the primary supports of the practice of psychotherapy. He termed it "a major scandal of our profession" (Stevenson, 1959). Later investigations, such as those of Brill et al., (1964), Cross (1964), and Phillip May (1969) reached similar conclusions. Ellsworth notes that, "While brief, time-limited psychotherapy may be helpful to patients and families (Schlien, 1957; Lorr et al. 1966), particularly at a time of crisis (Levy, 1966), it must be concluded that the value of prolonged psychotherapy remains undemonstrated" (Ellsworth, 1974). Yet the private model of treatment remains the method of choice in many, perhaps most, mental health settings throughout the United States.

Why this relatively ineffective private model of human intervention continues to thrive is a matter of speculation; however, reasons for its continuance must include such factors as the fact that practitioners tend to be legitimized by prestigious institutions (e.g., hospitals and universities), and that the specialized psychological language of the practitioner has been widely adopted by public institutions and the courts.

In addition, given the arguments and evidence developed by the Braginskys, Szasz, and Scheff, it is difficult to avoid the real possibility that this method of treatment and the language that accompanies it serves as the basis for certain kinds of social control. That is, the psychotherapeutic treatment system, not known for its effectiveness, first becomes a language used for the accusation of mental illness. It then serves as a causal labeling system, one having the consequence of perhaps permanently placing the labeled person on the receiving end of social prejudice.

Private models of treatment make certain assumptions about both the locus of a person's deviant behavior and its treatment. These assumptions distill to the essential point that the locus of deviance is within the person. Thus treatment strategies must be aimed at those problems inside the person. The net result has been a practitioner-defined set of mental health problems and remedial strategies. To the extent that the client has resources to influence the definition of his problem, such as purchasing the services of a private practitioner or a lawyer, he is likely to more effectively negotiate his place and status in a treatment system. To the extent that the client lacks these resources, he is likely to be at the mercy of the state, as embodied in the courts and public mental health treatment systems.

Belatedly, and perhaps reluctantly, mental health practitioners have begun to face the realities that have confronted other health and helping professions. That is, private treatment events are beginning to be constrained by a process of public review, and in mental health the social-behavioral value assumptions operating in treatment methods are being challenged. The historical absence of externally generated criteria for therapeutic outcomes, information feedback on those outcomes, and the presence of powerful but unstated value assumptions regarding appropriate behavior in social interaction are major fissures in the structure of psychodynamic treatment systems. Taken together, in a period of increasing pressure for accountability in the human services, these problems may sufficiently weaken the structure so that it will collapse under its own weighty concepts, or be supplanted by more effective treatment modalities.

GETTING OFF THE HORNS OF THE DILEMMA

If we are to satisfactorily lay to rest the charges that we inadvertently act as political agents for social control it is essential that our treatment models become more open, our treatment procedures more specifiable, and our treatment outcomes more observable. In short, treatment must become a set of empirical operations that are clearly articulated and evaluated.

Such a change does not mean that one discards what seem to be workable and helpful conceptual models for treatment. It does mean that whatever a practitioner's theoretical orientation, he should be able to logically and clearly translate his treatment principles into specific treatment behaviors. Further, these behaviors must be articulated in terms of particular treatment goals, and therapeutic progress be measured in terms of goal attainment. Demonstrable treatment effectiveness, not ideological correctness, then becomes the criterion by which a practitioner and client make decisions regarding therapeutic progress.

The components of such a treatment system are likely to stem from a behavior-analytic, rather than a psychodynamic conceptual framework. By psychodynamic we do not mean the Freudian system; rather, we mean any system that poses the major determinants of complex social behavior as resting principally "inside the head" of the client. These intra-psychic determinants are difficult to specify, much less observe, and seem likely to reflect the biases, past history, and verbal and intellectual facility of the practitioner rather than the problems of the client. They do not easily lend themselves to an open, observable, empirically evaluated treatment system.

Behavior analytic strategies of treatment, stemming from early research in operant (Watson, 1924) and classical (Pavlov, 1927) conditioning are based on the methodologies of an experimental tradition, and focus on specifiable, empirical events. They have the evidence of laboratory research behind them, and the advantages of an increasing array of technologies within them. The more modern translators of these systems of learning, such as Ayllon and Azrin (1968), Skinner (1953; 1971), Ullmann and Krasner (1965), Tharp and Wetzel (1969), and Wolpe (1958) have specified many of the core components of behaviorally based treatment systems. The basic structure of these strategies involves four basic and complex questions:

1. What is the problem?
2. What are baseline data regarding the problem?
3. What intervention procedures can be designed to affect this problem?
4. What are the outcomes of intervention efforts and how are they assessed?

Defining the Problem: The Essential Issue

The practitioner and client have a joint responsibility in defining the client's problem. While in a more intra-psychic treatment system the practitioner would have the sole responsibility for assessing and defining the client's problem; in a behavior-analytic system this is a joint responsibility, shared by the client, the practitioner, and other legitimately relevant parties.

The task of this group is to explore and define the behavioral characteristics of the alleged problem and the consequences of the behavior. In addition, this group, particularly the client and practitioner, must focus on the behavioral constraints placed on the client by the social systems in which he lives. The issue of whether the client should change his behavior or attempt to alter the contingencies which surround his behavior in social systems is likely to emerge as a major and highly salient problem.

It is precisely this problem that, in its legitimacy, provides a major point of departure for behavioral and psychodynamic treatment systems. It is quite possible that, using behavioral methods of treatment, the problem is no more likely to be decided in the direction of attempting to change the social systems. Rather, in a behavioral treatment system the issue is a legitimate one, and one in which the client makes the ultimate choice. The probability of addressing this issue in psychodynamic treatment is considerably lower.

The client and practitioner have the additional task of exploring the client's views of what is important in those areas of his life related to the problem behavior. In short, they must explore his values, sources of motivation, and reinforcing environmental contingencies. These then become key components in the design of treatment methods and the development of reinforcers contingent upon the "to be changed" behavior.

The important point is that the value systems of the *client*, not the practitioner, become paramount in defining the treatment relationship, and its goals and methods. It is exactly this problem that has come to plague psychodynamic treatment methods. However humanistically concerned a psychodynamic practitioner might be about his client, his basic treatment model, including its diagnostic categories and labels, is couched within the value systems of the mental health establishment. And, by extension, it contains the values of those segments of community socio-political systems that have interests in the maintenance of the existing order of social power and control.

Such has been the dilemma of mental health. Our treatment methods for deviance control have evolved from organized restraints, such as institutions, to largely verbal psychodynamic therapies (for the middle class at least), and have been joined with the chemical controls of drugs. All of these have been based on the ultimate knowledge and decision-making wisdom of the practitioner. Despite the lack of evidence that these methods, with the possible exception of drugs, have been effective treatment methods, they have continued as the primary methods for deviance control. Not only must one wonder why, but it is difficult to avoid the conclusion of the Braginskys, that these psychological assessment and treatment methods perpetuate a system of socio-political control. This dilemma, particularly as it manifests itself at the onset of the treatment process, cannot be avoided by any treatment system. But it can be explored and resolved to the client's satisfaction if the practitioner and client actively incorporate it in their problem assessment and early design of treatment strategies. To the extent that these strategies are clearly stated, mutually defined, and behaviorally specified, the dilemma can be converted into a real and productive issue affecting treatment methods and outcomes. To the extent that the practitioner's values remain unstated and imbedded within his treatment model, the historical dilemma of the exercise of both social control and individual rights will continue and hinder effective human learning and change.

The Issue of Baseline Data

The dilemma of mental health treatment as a set of methods aimed more at social control than adaptive behavior change also arises in terms of the data used to establish a base from which change is measured. If one operates within a psychodynamic model, the baseline data are likely to be generated within an intra-psychic conceptual framework. Thus, the possible presence of a client's unstated intra-psychic conflict may be diagnosed as the problem and the target of treatment methods. Assessment procedures are likely to be organized in terms of the client's willingness to talk about this particular area of his life, articulate the conflict, etc. The behavior of the therapist will be centered around the generation of client verbal behavior related to the area of conflict.

The extent to which the intra-psychically oriented therapist is selectively attending to client behaviors that are related to the values of the therapist and his treatment model then becomes a major issue. If this is what is actually happening, as we believe it likely to be, another step has been taken in maintaining the treatment/social control dilemma.

To those readers who are familiar with a behavioral orientation towards gathering baseline data, it should not be necessary to elaborate these procedures. To those readers who are not familiar with these procedures, suffice to say that there is an attempt to quantify the occurrence of the problem behavior and the situations in which it occurs, and to examine controlling stimuli related to the onset and termination of the behavior. Functional relationships are then established between these variables, and behavioral changes noted as later intervention procedures are carried out and the behavior is brought under new patterns of stimulus control.

Intervention Procedures

At risk of overstating what at this point may be the obvious, psychodynamic and behavioral methods dramatically and radically differ in their treatment operations. The value assumptions which guide the traditional therapist's behavior in treatment are likely to maintain the legitimacy of certain social class related statements from the client, and implicitly declare the illegitimacy of others. To a large extent these treatment methods are aimed at verbal behavior only, and perhaps because of this do minimal damage. But, neither are they likely to result in treatment gains for, say, the local mechanic who lacks sophisticated verbal skills and who is asking for help from a mental health agency.

By contrast, behavioral treatment methods focus on the problem behavior itself and the stimuli which produce and maintain that behavior. An attempt is made to restructure contingencies so that client behavior will change in ways that the client has defined as meaningful. The extent to which the particular kinds of behavior change being generated are social-class related is a function of the choice the client has made in defining his problem, not the value assumptions contained in a practitioner's diagnosis of the problem.

Outcome Assessment and Evaluation

The termination of psychodynamic therapies is a complex problem. This complexity stems from the fact that in assessing therapeutic outcomes psychodynamically-oriented therapists want, in effect, to have their cake and eat it too. That is, in the formal training of practitioners who are psychological determinists, there are certain logically developed criteria which signal an appropriate time to terminate therapy (e.g., a breaking of dependency upon the therapist, being able to talk about certain areas of life without the occurrence of previously present anxieties, etc.).

However, traditional therapists are also willing to accept the occurrence of behavior outside their formal logical model as data supporting successful psychotherapy. For example, a client may simply not reappear for therapeutic interviews. The practitioner may hypothesize that this signals independence from the therapeutic relationships and thus may be viewed as a successful therapeutic outcome. This after the fact reasoning is not without some legitimacy, for people often do terminate therapy when they, not the therapist, feel it is no longer warranted.

However, people also terminate therapy and then commit suicide, overdose on drugs, beat their children, and engage in a host of other negative behaviors. If the outcome evaluation game is to be played fairly, then these cases should also be included in a practitioner's assessment of his efforts. Too often they are not.

It is also likely, given the long-term nature of many psychodynamic treatment strategies, that various processes in the therapeutic relationship become ends themselves. Positive regard, transference, countertransference, and various other concepts related to the client's emission of certain verbal behaviors become important goals. As such they become the criteria by which the therapist views the effectiveness of his treatment efforts; criteria which may supplant the real problem of changing behavior in the client's everyday life. In addition, the practitioner, particularly the private practitioner, is likely to be the sole agent for evaluating his client's therapeutic outcomes. Thus, a researcher might argue that the fox is left to guard the henhouse.

The problems of outcome evaluation in behavioral treatment methods contain many of the measurement complexities occurring whenever one attempts to measure changes in human behavior and attribute those changes to a particular intervention strategy. However, behavioral treatment methods, having specifiable procedures and client-practitioner validated outcomes, lend themselves more easily and readily to quantifiable outcome measures. That is, the behavior in question did or did not change. If it did, the changes can be specified, and the effectiveness of treatment methodologies evaluated. Further, specific analyses of treatment methods or components of those methods can be examined in terms of costs, and costs per behavior change (e.g., Howley, 1973; Foreyt et al., 1973).

Another strength of behavioral treatment methods lies in the fact that outcomes are mutually defined by the client and the practitioner, rendering outcome assessment and evaluation a relatively public event. In larger treatment systems, such as clinics and mental hospitals, outcomes might be further reviewed in a broader public context. This characteristic of public review contrasts markedly with the private review of psychodynamic methods.

ARE WE OFF?

Completely removing the methods and delivery system in mental health from the dilemmas posed by their being joined with the issue of social control is probably an impossibility. Life is too complex, and our own lives as well as those of our clients are perhaps inextricably related to community value systems and the related problems of social control. But, the dilemma posed by the Braginskys and the dangers discussed by Levine can be minimized. As a step in this direction we would suggest that behaviorally based systems of treatment be more carefully considered, particularly in those training programs that claim to equip some people to help other people deal with their problems. These treatment methods have numerous advantages, among them the following:

1. Personal values, including those of both therapist and client, are more likely to be articulated at the outset of treatment. They are unlikely to remain unstated, yet powerful influences in guiding the behavior of the parties engaged in treatment procedures.

2. The client's problem is stated in terms of specifiable behaviors, which are jointly defined by the client and practitioner.
3. Treatment procedures are behaviorally specified and designed to lead to behaviorally specified treatment outcomes.
4. The above outcomes are subject to review by the client, the practitioner, and other legitimate parties from "the community."
5. The practitioner can be held accountable for his professional efforts, as can treatment delivery systems.
6. Specific analyses of treatment methods, the organization of these methods in delivery systems, and outcomes become possible. These analyses, such as cost and cost-effectiveness analyses, can serve to provide information for feedback loops to the practitioner and the supporting organization. This form of information feedback is likely to lead to improvements in the quality of the delivery system.
7. There is a growing body of empirically based evidence linking the effectiveness of particular behavioral treatment methods to specific kinds of problems. Thus, for the practitioner there is a growing fund of data-based information on which treatment decisions can be made.
8. On the basis of the above data, it will become increasingly possible to train effective practitioners without the necessity of their spending many years in graduate training. A cadre of competent sub-doctoral professionals can become a reality, not just a promise.

For hundreds of years the problems of managing deviants and protecting the public interest have remained largely unsolved. Then and now, institutions have been constructed to house and hold people who violated the norms of community life. In the recent past ineffective treatment methods have arisen, creating the illusion of treatment. This has been particularly true in the operation of large public mental hospitals. Although more subtly managed, it has also been a problem in private treatment systems. Overall, research would suggest that, when treated with these illusory methods, the client probably fares no better or no worse than if there were no treatment at all. Whatever behavior changes do occur are probably a function of the social environment in which the patient lives.

We may be entering a new era in mental health treatment methods. The data now available indicate that these methods are far more effective than earlier treatment, but the history of mental health treatment would remind us that unless we view our developing methods with some skepticism, and apply a rigorous evaluative research strategy to any judgments of their effectiveness, we run the risk of repeating the errors of the past.

Professionals in mental health services have had a distinct inclination to substitute good intentions and belief in the efficacy of treatment methods for outcome evaluation. The logical bases for decisions regarding treatment services have more closely resembled those of religion than science. It is doubtful that the success rate has been higher.

As long as the mystique of mental health treatment remains, as long as the practitioners retain their priestly roles and specialized languages, and as long as the fiction that we can't effectively measure treatment outcomes continues to flourish, we will continue to function as society's agents of control. The dilemma will remain, for the opportunity to demythologize the treatment process, and make it a set of behaviorally specifiable human relationships will have been missed.

REFERENCES

Ayllon, T. & Azrin, N. *The Token Economy: A Motivational System for Therapy and Rehabilitation.* New York: Appleton-Century-Crofts, 1968.

Braginsky, D. D., & Braginsky, B. M. "Stimulus/Response: Psychologists: High priests of the middle class." *Psychology Today,* December, 1973, 7(7), 15-142.

Brill, N. Q., Koegler, R. R., Epstein, L. J., & Forgy, E. W. "Control study of psychiatric outpatient treatment." *Archives of General Psychiatry,* 1964, 10, 581-595.

Cross, H. J. "The outcome of psychotherapy: A selection analysis of research findings." *Journal of Consulting Psychology,* 1964, 28, 413-417.

Ellsworth, R. "Consumer feedback in measuring the effectiveness of mental health programs." In Guttentag and Struening (Eds.) *Handbook of Evaluation Research,* in press.

Foreyt, J. P., Davis, J. C., Rockwood, C. E., Desvousges, Wm. H., & Hollingworth, R. "Cost benefit analysis of a token economy program." Paper presented at annual meeting of the Southeastern Psychological Association, April 6, 1973.

Howley, T. *Cost Analysis of an Institutional Token Motivational System.* Unpublished Masters Thesis, Drake University, 1973.

Kenniston, K. "How community mental health stamped out the riots (1968-78)." *Transaction,* July/August 1968, 21-29.

Levine, D. "The Dangers of Social Action." In D. Harshbarger and R. Maley (Eds.) *Behavior Analysis and Systems Analysis: An Integrative Approach to Mental Health Programs.* Kalamazoo, Michigan: Behaviordelia, 1974.

Levy, R. A. "Six-session outpatient therapy." *Hospital and Community Psychiatry*, 1966, 17, 340-343.

Lorr, M., McNair, D. M., & Goldstein, A. P. "A comparison of time limited and time unlimited psychotherapy." Cited in Goldstein et al., *Psychotherapy and the Psychology of Behavior Change.* New York: Wiley, 1966.

May, P. R. A. *Treatment of Schizophrenia.* New York: Science, 1969.

May, R. *Love and Will.* New York: Norton, 1969.

May, R. *Power and Innocence.* New York: Norton, 1972.

Pavlov, I. P. *Conditioned Reflexes.* New York: Oxford University Press, 1927.

Schlien, J. M. "Time-limited psychotherapy: An experimental investigation of practical values and theoretical implications." *Journal of Consulting Psychology*, 1957, 4, 318-322.

Scheff, T. J. *Being Mentally Ill: A Sociological Theory.* Chicago: Aldine, 1966.

Skinner, B. F. *Science and Human Behavior.* New York: Macmillan, 1953.

Skinner, B. F. *Beyond Freedom and Dignity.* New York: Knopf, 1971.

Stevenson, I. "The challenge of results in psychotherapy." *American Journal of Psychiatry*, 1959, 116, 120-123.

Szasz. T. S. *The Myth of Mental Illness.* New York: Hoeber-Harper, 1961.

Tharp, R. G., & Wetzel, R. J. *Behavior Modification in the Natural Environment.* New York: Academic Press, 1969.

Ullmann, L. & Krasner, L. (Eds.) *Case Studies in Behavior Modification.* New York: Holt, Rinehart, and Winston, 1965.

Watson, J. B. *Behaviorism.* Chicago: University of Chicago Press, 1924.

Wolpe, J. *Psychotherapy by Reciprocal Inhibition.* Stanford, California: Stanford University Press, 1958.

It is likely that our behavior as participants in human service delivery systems is, in large part, a function of the problem definitions and resulting delivery system goals and contingencies which we have developed over many years. Concomitantly, in mental health an elaborate language of problem definitions and problem solving strategies has emerged. It is this language, Ford suggests, that has placed blinders on our professional vision, and limiting constraints on our professional services.

The continued conceptual separation of mental health and mental illness, and deviance and non-deviance, in human service delivery systems is likely to have far reaching and, Ford suggests, very dysfunctional consequences for society. Conceptually integrative approaches to human development and the design of human services are critical problems, problems that need to be dealt with now. To continue our historically categorical approaches to the solution of human problems will unnecessarily burden consumers and care-givers with marginally effective problem solving strategies. It is toward the alteration of this condition that Ford addresses himself in the following paper.

Donald H. Ford

CHAPTER 3

Mental Health and Human Development: An Analysis of a Dilemma

THE PRESENT HUMAN SERVICES DILEMMA

This century has been marked by a growing interest in, concern for, and sophistication about individual and social health within our culture. As a result of this concern, economic investments have become huge and are still growing, as illustrated by the billions of public and private dollars spent on programs with labels like mental health, mental retardation, family services, child development, corrections, rehabilitation, gerontological services, welfare services, manpower development, youth services, treatment of alcohol and drug addictions, law enforcement, community development, model cities, or regional planning. More labels could be added to that list, but these are sufficient to demonstrate the multiple and overlapping efforts that have been growing within our society to help people cope more effectively with their difficulties of living and to improve their quality of life.

Our society has arrived at a point in time where the huge costs of this total social effort and the questionable or undocumented effectiveness of many programs are leading to a demand for reassessment and reorganization to try to produce more effective efforts at lower cost. This concern is reflected all the way from the highest levels of the federal government to local communities, as well as among professionals in these fields. For example, in his recent review, Cowen (1973) made the following summary statement.

> Cold evaluation of past dominant MH approaches invites the criticisms that as articulated: they have not provided the manpower and resources to cope with evident problems and latent needs; they have not solved

the baffling problems of major mental illness; their pivotal techniques have had, at best, limited effectiveness; and they either do not reach, or are inimical to, major segments of the population urgently requiring help. To those chronic concerns we must now add the new challenges posed by explosive social problems such as poverty and violence and their links to disorder, not seriously seen before as problems for the mental health fields.

Mental health is one of an array of labels referring to services for different clusters of biological, psychological, and social concerns. To create completely separate services for each category of concern (e.g., criminal offenders, delinquent youth, emotionally disturbed, physically handicapped, mentally retarded, disrupted families), has been the pattern of the past but probably will not be the pattern of the future. Congressional legislation in fields such as developmental disabilities, social services, mental health, and rehabilitation in past Congresses, as well as the present one, moves strongly toward the recognition of the interrelatedness of human concerns and the need to develop correlated patterns of human services. Moreover, as scientists and professionals in this field, we know that this fragmented, categorical approach should not be the approach of the future because it is contradictory to our understanding of how people develop and live. We must develop approaches which serve persons in their social context rather than treating pieces of people as symptom categories, and too often leaving consumers stranded in the No Man's Land of interagency jurisdictional problems.

Therefore, this paper is based on the assumption that a fundamental review of both the concept of mental health and its implementation is not only timely, but also essential. My reading of the present social and political signs suggests that social policy, as well as financial and public support for programs labeled "mental health", are not growing and may well be declining (e.g., APA *Monitor*, April, 1973). This past year, the President of the American Psychiatric Association, Perry Kington, called a second joint Commission on Mental Illness and Health to renew the momentum produced by the first one a decade ago. Social policies and programs vaguely defined, whose social impact is not clear or obvious, and for which there is not a strong and organized constituency, other than the people whose jobs depend upon the existence of the program, cannot expect to maintain or increase public support. There is a powerful and justifiable current of concern in this country about human services which can be ignored by mental health professionals only at grave risk to the programs they believe valuable.

THE CONCEPT OF MENTAL HEALTH

To what domain does the concept of mental health refer? The *Encyclopedia Britannica* states:

> Mental health cannot be defined with precision, since it is closely related to the customs and requirements of society. Since customs vary and societies differ, there is a multiplicity of variable factors . . . the physician is apt to think of mental health as the absence of mental illness, and such definitions serve his purpose. Mental health in the broader sense suggests a degree of happiness and satisfaction, under conditions that warrant such a state of mind and a capacity for making satisfactory personal and social relationships.

This difficulty in defining the domain of mental health is not new. For example, in the early 1950's Ginsberg (1955) stated:

> One of the great theoretical lacks in mental hygiene activity seems to me to be that we do not have an adequate definition of mental health . . . this inevitably results in a confusion of goals and an uncertainty of means; it creates a situation like that of a hunter stalking an unknown prey with weapons which may turn out to be quite unsuitable. The notion of normality . . . is based to a large degree and often solely on the value system of the author using the term; . . . it stems also from a lack of exact and scientifically determined criteria for mental health, which is even more serious. Early attempts to define mental health simply equated it with an absence of mental illness. This largely begged the question, since, except for evidences of the most serious disease, there is no satisfactory definition of mental illness. The border line between mental health and illness is vague. It even fluctuates from one cultural, socio-economic setting to the next.

Since experts cannot define it, it is not surprising that the public is uncertain about what mental health is and therefore finds it difficult to relate evidence or personal experience to document the value of such programs. Without that, public support for any program declines and along with public support, funding declines.

A domain that is undefinable is a pseudo-domain and has little utility for scientific and social purposes. A domain which is undefinable does not identify a set of issues on which to focus, a knowledge base upon which to build, a set of methods which are applicable, the kind of training essential for dealing with the issues, the questions to be addressed, or the public understanding which is essential. Is it possible that our manner of thinking about this whole problem is part of our problem?

Is it possible that the terminology and conceptual representation of "mental health" presently used is one of our barriers?

The purpose of this paper is to explore the domain of mental health, formulate some views about it, and consider the implications about those views for the future. It should be recognized from the beginning that value judgements are involved and will affect the argument presented. I will first examine some of the ways in which the present nomenclature may be seriously misleading, and then I will consider some factors which I believe should guide our thinking and approaches in the future.

Weaknesses of the Concept

There are at least four basic weaknesses inherent in the nomenclature of "mental health".

Conceptual Inadequacies

I suggest that the label "mental health" has become an anachronism, reflecting an unacceptable theoretical and scientific position. Webster's *New World Dictionary* refers to "mental" as meaning something done by or carried on in the mind. The World Health Organization now defines the term "health" to include complete physical, mental and social well-being, broadening it greatly from its primary original reference to the biological integrity and effective biological functioning of the body.

The original use of the term "mental illness" was to refer to human dysfunctions represented by identifiable types of behavior such as hallucinations, delusions, loss of memory, and unresponsiveness to social stimuli which seemed to be the product of dysfunctions of the "mind" which controlled imaginal and sensory responsiveness. We've now come to realize that the term "illness", used to denote a disease process or a biological dysfunction, does not appropriately apply to most disruptions of psychological or behavioral functioning in either an etiological or functional sense.

The term "mental health" does not accurately reflect our current understanding of human behavior. The science of psychology, like other sciences, has approached its phenomenon of concern analytically. It has sought to take human beings apart to see what makes them tick. The same may be said of biology. Therefore, the basic research literature has developed an extensive knowledge base about various behavioral and biological response processes such as perception, sensory functioning, psychophysiology, endocrinology, cognitive functions, biochemistry, memory, emotions, electrophysiology, attitudes and opinions, learning, motivation, and neuropsychology. Such fundamental scientific work is crucial to the development of the knowledge base necessary to understand ourselves. But, what we increasingly are realizing is that a person functions as a unit and these various response processes, while appropriately studied as sub-elements of a person, always function in highly integrated patterns of interaction with one another.

We are coming to understand the way emotions and their biochemical correlates may disrupt thinking or modes of performance. We are beginning to understand the way attentional processes may filter sensory inputs and control cognitive processes. We are beginning to understand the way thoughts may coordinate or disrupt other behaviors. We are beginning to understand the way physical restrictions may affect psychological processes. In other words, we are beginning to understand that we cannot treat man as a mind or a heart or a mouth or a body, but that for any practical purpose we must treat him as a person; as a highly complex unit, with biological, psychological, and social characteristics always occurring in complex interrelated patterns in some environmental context.

The term "mental health" perpetuates a mind-body dichotomy which is inaccurate and fundamentally misleading. Yet, thinking in those terms still permeates much of our terminology. As our understanding of the ways in which our patterns of linguistic notation powerfully influence the structure of our knowledge and activity, has grown (Pike, 1967), the handicap of inappropriate terminology has become more obvious.

Programmatic Inadequacies

The term "mental health" does not clearly identify categories of behavioral or social problems with which our society is concerned. Is a juvenile delinquent or a drug addict, a criminal or an alcoholic, a problem of mental health? Are child beating or marital discord, racial discrimination or sex typing, insomnia or ulcers, prostitution or homosexuality, police brutality or employee absenteeism, loneliness or agression, youthful dropping out or aging, problems of mental health? Mental health professionals seem to believe they are. Long ago, Wechsler (1930) noted:

> Enthusiastic mental hygiene tells us that it is concerned with the prevention of mental deficiency, criminality, the psychoneuroses,

> the psychoses, anti-social traits, family unhappiness, divorce, prostitution, alcoholism, sexual perversion, epilepsy and other such simple matters.

A recent analysis of publications in the first seven years of the *Community Mental Health Journal* (Cowen, 1973), Golann's, (1969) *Coordinate Index Reference Guide to Community Mental Health*, and the National Institute of Mental Health *Mental Health Program Reports No. 2* (1968) all illustrate that the label, "mental health", continues to be used to cover a huge diversity of human concerns. A delinquent or a criminal offender is usually dealt with by a Bureau of Corrections (although such a person may be mentally retarded, physically handicapped, emotionally disturbed, etc.). Family services for the poor are associated with welfare programs. Programs to encourage improved family nutrition and family management are provided through the Department of Agriculture. I could list more but these are sufficient to make it clear that the term "mental health" does not mean to the public, to other professionals, or to the government, what mental health professionals have sought to make it mean to them.

Methodological Inadequacies

The term "mental health" does not identify a category of validated methods, technologies or intervention strategies peculiar to its domain. While verbal psychotherapy might be thought of as such a technology, verbal interview methods are widely used to help people understand themselves better and to help them try to alter their behavior. They are used by clergymen, lawyers, salesmen, teachers, labor leaders, executives, and parents to name a few. The same could be said for methods of group processes or intervention with drugs, or teaching or training approaches, or consultation methods. Unlike medical care with which the public can readily identify certain intervention methods, (e.g., surgery, medication), mental health has no such clear identity.

Professional and Knowledge Foundation Inadequacies

The term "mental health" is discipline and guild-related (and to some extent, discipline and guild-controlled), and therefore, does not provide a satisfactory frame for essential integration across disciplines and professions. Work with mental illness (and its derivative, mental health) has been controlled by physicians, more lately called psychiatrists. Psychologists fought their way into the domain after World War II. Psychiatric social workers, and psychiatric nurses have sought their portion of the action. It is interesting to examine the bibliography of various reviews of mental health research or mental health programs (e.g., Cowen, 1973; Ford and Urban, 1967). Most of the references are from the psychological or psychiatric literature. Despite the fact that the term "community" has been tacked on to the term "mental health", one seldom sees references from anthropology, sociology, political science, community development, and the like.

Helping distressed people try to overcome their unhappiness is a big industry in this country and provides a very large job market for psychiatrists and psychologists. Thus, the knowledge base, the methods, and the economics are dominated and heavily controlled by a small array of disciplines and professional guilds. There are a large variety of other professional guilds developing around and promoting the development of approaches for dealing with special categories of human problems which the public does not identify with mental health (though mental healthers might like it to).

We are all aware of the huge sums of money our society is pouring into the field of criminal justice now; and the field of corrections is developing its own professional training programs, its own guild structures, and its own program controls. Mental retardation programs have a powerful and dedicated public constituency behind them, and they too are developing their own programs, their own guild characteristics, and their own patterns of guild control. Social workers still control services to the poor despite the heavy attacks upon their accomplishments during the last decade. A large domain of gerontological services is rapidly developing and there are signs of the development of a gerontological guild to control those services in that job market. More examples could be listed but these are sufficiently to the point.

Human beings have been broken into pieces and divided among various professional guilds, each serving a special interest. It is being increasingly recognized in both legislative and professional circles that this is no way to treat human beings. An increasingly highly-educated public will most probably begin to show a growing recognition that it is not in their personal or economic interest to have it done that way.

Summary

This section seeks to outline the argument that the nomenclature of "mental health" is inappropriate and misleading conceptually, programmatically, methodologically, and professionally. This does not mean that much of the work being conducted under that nomenclature is not valuable, nor that the professionals involved are not dedicated, competent people, by and large. On the contrary, I would argue that the approaches which have been and are being developed under the rubric of mental health have potential applicability over a wide spectrum of problems, and that our socially significant and humane objective should be to seek to create correlation and syntheses of these developments with developments in other human service fields and other foundation disciplines. That is the direction in which our growing understanding of ourselves and our relationships to our environments is leading us. That is the direction in which social policy is moving at national levels. That is the direction in which economic necessity is driving us.

We cannot remain encapsulated in the confusing and overlapping categorical approaches to human problems now dominating our country. The guild walls must come down. The turf tariffs must be eased so that a common market in human services might have an opportunity to evolve. We should not, and perhaps cannot, remain loyal to old categories and nomenclatures because of their past utility or the present job security they appear to provide. Competent and conscientious people with valid theories and effective methodologies do not need the security of old nomenclatures or old delivery systems to protect their future. A new correlation of effort, a new theoretical and methodological synthesis, will probably require a new or revised conceptual frame and nomenclature which is less categorical in nature than that which presently dominates us.

ASSUMPTIONS TO GUIDE DEVELOPMENT OF ALTERNATIVE APPROACHES

The first section of this paper has tried to identify some of the difficulties associated with the continued use of mental health as a primary nomenclature for trying to deal with the wide range of human concerns for which some seek to make it the umbrella concept. In this second section an attempt will be made to identify some guidelines for developing alternative approaches. I am very much aware of the hazards of trying to comment upon an issue as broad and fundamental as this one in the brief time possible in this paper. However, I believe we must all seek alternatives which will represent improvements. Therefore, I will set forth as a series of assumptions some propositions which I believe might guide us. My suggestions stem from an attitude set forth in Garrett Hardin's (1968) article, "The Tragedy of Commons".

> It is one of the peculiarities in the warfare between reform and the status quo that it is thoughtlessly governed by a double standard. Whenever a major reform is proposed it is often defeated when its opponents triumphantly discover a flaw in it. As Kingsley Davis has pointed out, worshipers of the status quo sometimes imply that no reform is possible without unanimous agreement, an implication contrary to historical fact. As nearly as I can make out, automatic rejection of proposed reforms is based on one of two unconscious assumptions: 1) That the status quo is perfect; or 2) That the choice we face is between reform and no action; if the proposed reform is imperfect, we presumably should take no action at all, while we wait for a perfect proposal.
>
> But we can never do nothing. That which we have done for thousands of years is also action. It also produces evils. Once we are aware that the status quo is action, we can then compare its discoverable advantages and disadvantages with the predicted advantages and disadvantages of the proposed reform, discounting as best we can for our lack of experience. On the basis of such comparison we can make a rational decision which will not involve the unworkable assumption that only perfect systems are tolerable.

I suggest the following propositions might guide us in designing constructive reforms to our present approaches. The propositions are not all of the same order, (e.g., some are value assertions; others are programmatic statements).

Assumption I

A fundamental social value guiding our society is an emphasis on the prime importance of the individual's fullest and most creative development and living. Our society's objective is not just on keeping people alive, but also upon helping them to live creatively and with quality. Our society's effort to create great social institutions is not intended to make the individual subservient to the institution but rather to create and continually transform those institutions to more fully serve and enrich individual life. Of course, in a mass society, individuals must have institutional arrangements which enable them to live together harmoniously as a group while developing creatively as individuals and, therefore, a continuing dynamic tension exists between the functioning of individuals and institutions. Whatever arrangements and approaches are developed to cultivate effective human services should be designed to recognize and serve this fundamental social value of our culture.

Assumption II

A second assumption is that one of the conditions essential to the achievement of creative individual development and living is the fullest cultivation possible of the characteristics of self determination, self direction and self control. To say it differently, we seek to develop a society in which individuals are in as full control of their own lives as possible, modified only by the essential need for cooperation and compromise which is basic in all forms of group social living. This is not only a social value but also reflects a basic theoretical assumption for some who take an open systems approach to understanding the functioning of the individual. This assumption has certain important implications for mental health and human services. Too frequently, in our zeal to apply the division of labor principle, we have increasingly led people to believe that they can't resolve their own difficulties, but must seek the help of an "expert" in all aspects of their lives. The social strategy implicit in the substance of many of our human services involves experts in doing things to and for people rather than relying more heavily on people's capacity to guide their own lives. Within the framework of this assumption, a central focus should be on creating conditions under which people can do for themselves.

We should constantly seek to transform our technical expertise into a form that the ordinary citizen can use. One example will illustrate this point. A considerable number of children develop social-emotional (i.e., mental health) disturbances. A typical pattern has been for parents to take the child to an "expert" to get "cured"; and the psychiatrist, clinical psychologist, or social worker may give the child "therapy" at the "clinic". Not only is this an expensive procedure, but also it does not give adequate attention to the child's social context. Bernard and Louise Guerney (1969), and colleagues in Penn State's College of Human Development, have spent several years on research demonstrating that parents can be trained to give "psychotherapy" to their own children in their own homes for many such disturbances with results as good or better than those achieved by "experts" at the "clinic". Moreover, this approach enables parents to become more effective in dealing with other children as well. It is simultaneously remedial for one child and preventive of future difficulties for other members of the family. Behavior therapists are experimenting with analogous approaches, as are others.

There are probably many ways which could be developed in health, welfare, and social services in which the concept of training people to do for themselves rather than doing for or to them could be applied, and we should seek to design our services to focus on that social strategy. That is the best long-term approach to reducing the costs of human services and to cultivating individual freedom. The growing educational level of people in our society is making such approaches increasingly feasible.

Assumption III

A sound approach in our culture to the evolution of effective human services should focus on developmental and preventive, as well as remedial approaches. This is a simple yet powerful idea. It is more humane and less costly to create conditions under which people can live well, stay healthy and out of trouble, than it is to wait until they are in a mess and then try to fix things up. Figure 1 represents the evolution in the focus of human services in this country. The first primary focus was on remediation, (i.e., when people get in trouble try to help them recover). In the early 1900's, the preventive focus began to gain real momentum. About the middle of the century the most fundamental objective began to become more visible, (i.e., the fullest positive development of people's capabilities). Developing one's potential is more than prevention. The World Health Organization says that health is a state of complete physical, mental, and social well being and not merely the absence of disease or infirmities. Two examples will illustrate the power of this idea.

FIGURE 1

The Evolution in Focus of Human Services in Different Periods, From Remediation, Through Prevention, to Positive Development

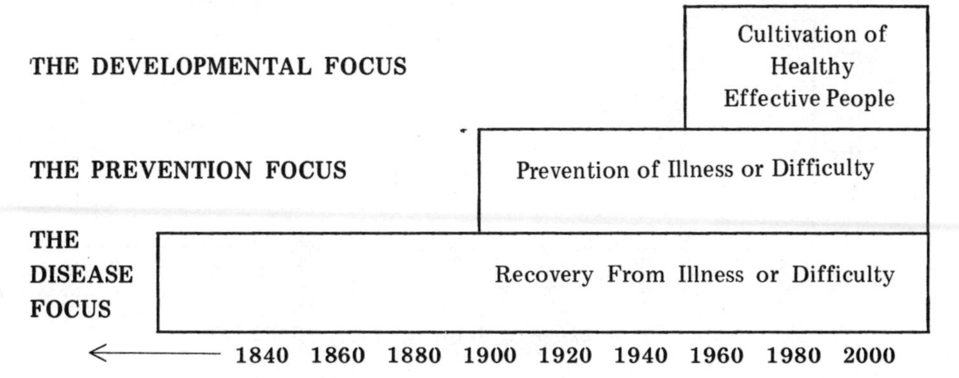

FIGURE 2

Reduction of Typhoid Fever in Philadelphia Following Treatment of the Water Supply

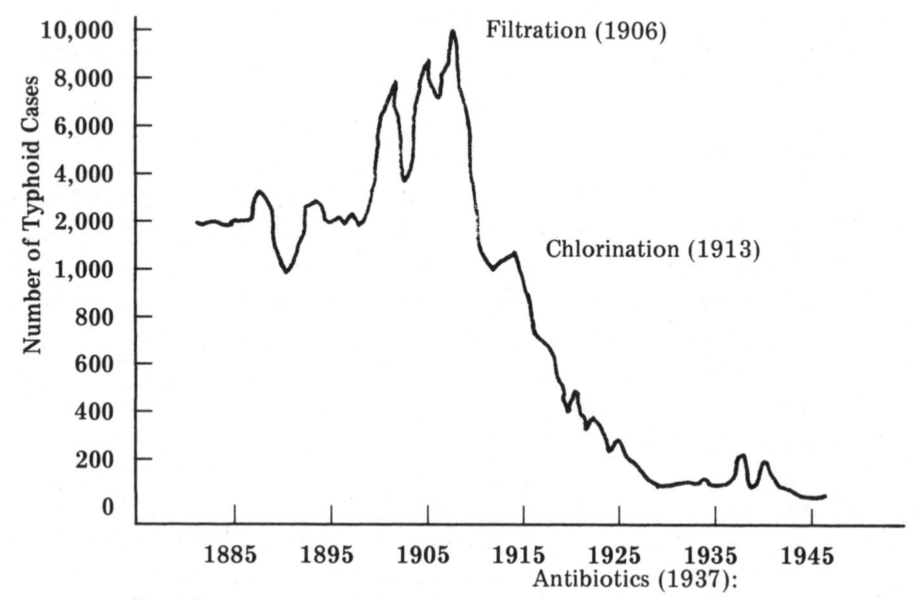

(Data Source: Dr. Angelo M. Perrl)

FIGURE 3

Mortality Per 10^5 Living in England and Wales 1851 — 1935

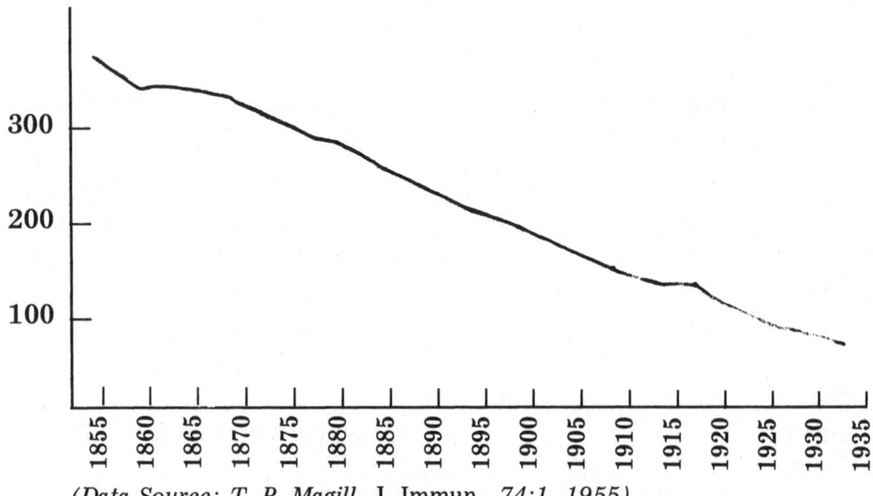

(Data Source: T. P. Magill, J. Immun., *74:1, 1955)*

Figure 2 shows that eight to ten thousand typhoid fever cases occurred in Philadelphia at the turn of the century. Twenty-five years later the problem was almost eliminated. How? Not by the medical treatment of individuals, but by the preventive approach of purifying the water, first through filtration in 1906, and then chlorination in 1913. Remedial treatment could never have eliminated the problem — it could only alleviate the pain.

Figure 3 illustrates the steady and dramatic decline in the crude death rate between 1855 and 1935. Why? A minor portion may be attributed to improvement in medical treatments. Some may be attributed to preventive measures such as immunization. However, the industrial revolution, with the general and progressive improvement in the quality of living it produced, was probably the major factor. Warmer and cleaner homes, better housing, more, and higher quality food supplies, purer water, better waste disposal, less exhausting work schedules, better child care and similar factors were probably far more influential on the death rate than the medical advances of that period. There has been a huge drop in mortality rates per one hundred thousand at all age levels. One consequence of this change is that whole new human concerns that never existed in man's history, exist today, such as those resulting from the rapidly growing and very large population of elderly people.

The blunt fact is that the natural scientist, agriculturalists, and engineers have probably had a far greater impact on keeping us all alive and healthy longer than all the physicians in the world. Don't mistake my meaning. I think we need physicians, and good ones. But, they are our emergency back-up system. Our greatest positive results for the least cost come from cultivating constructive development, quality living conditions, rather than relying primarily on remedial measures. In his recent review on social and community interventions, Cowen (1973) asserts that the remedial assumption has implicitly dominated the mental health professions' views about how best to approach and deal with complex problems of disordered behavior.

> Clinical theorizing, derivative practice, and research all mirror the implicit assumption that the time to initiate help is when a person experiencing problems, sometimes grave, recognizes them and/or is rendered ineffectual by them. At such moments, he, or those responsible for him, send out a cry for help. We have thus been markedly shaped by an end state mentality wherein interventions begin when routine dysfunctions (e.g., neuroses, addiction, schizophrenia) exceed threshold, disrupt the individual's happiness or effectiveness, or intrude on others. The core

aim of ensuing interventions has been to relieve the debilitating conditions, (i.e., to rehabilitate an already disabled — sometimes chronically so — individual!)

Assumption IV

Both our social orientation and our present scientific knowledge of ourselves indicate that we must design approaches which deal with the person as an integrated bio-psycho-social unit, functioning in systematic patterns, who must be dealt with in those terms and not as a collection of parts with each part susceptible to intervention and modification separate from the others.

All of the many human functions must take place in coordinated fashion if one is to have a *person*, a unitary being, striving to do something with life. A disruption of one function will affect others. A lack of goals will result in erratic, uncoordinated, or undirected behavior. Poor planning may produce misdirected behavior. A physical handicap, such as paralysis, may prohibit certain essential activities. A sensory deficit, such as blindness, will limit capability of carrying out environmental transactions, information feedback, and what is learned. Intense emotion may destroy judgement or coordination. Faulty self evaluations may distort interpretation of one's behavior. Biochemical dysfunctions may influence food processing or emotional functioning. A person can be understood only as a complex organism, functioning in a unitary way in transaction with an environment which limits and facilitates what is feasible for a person.

For example, a man develops a bleeding ulcer. With a controlled hospital environment, bed rest, medication and a special diet, the ulcer is beginning to heal and he is sent home. Six months later he is back with the same problem. Why? He drinks fifteen cups of coffee and smokes three and a half packs of cigarettes a day. He has a problem daughter he can't control, and a wife who is a shrew. His ulcer is a product of a biological vulnerability, faulty habit patterns, and a mess in his family life which he is partly responsible for creating. His cure requires more than physical medicine and will require other types of professional help in addition to medical specialists.

To implement this point of view, it becomes essential, therefore, that the knowledge base for this approach be one which draws together what is known from a variety of disciplines and professions of which psychology is one, but not the only one. It suggests that arrangements must be sought in both research and service settings to produce this kind of integrative approach, both in terms of knowledge and in terms of intervention strategies.

Assumption V

It must be fully recognized that behavior is always a function of a transaction within an environment and can be understood and dealt with only in those terms. A standard principle presented in introductory psychology texts for years has been that behavior is a function of the situations under which it occurs. However, in practice, far more emphasis has been given to the behavioral, rather than the situational, side of the equation. Yet, as our knowledge has increased through research ranging from studies of sensory deprivation to the social impact of poverty, the interdependence of behavior and situational factors has become increasingly clear. Time after time psychotherapists have discovered how rapidly the puny effects of their efforts to effect individual behavior change in private interviews in their offices are washed away in the currents of the individual's natural social environment. A variety of training methods, including behavior modification approaches, have been demonstrated to produce behavior changes in closely controlled environments. But, there is also evidence that those behavior changes frequently melt away when the individual is returned to the powerful forces in his or her natural, everyday environment.

It follows from this assumption that interventions focused on situational or environmental variables, rather than directly on behavioral variables, should have major preventive and developmental potential. In his recent review, Cowen (1973) argues that social and community intervention approaches offer a genuine alternative to traditional "mental health" approaches.

> It is active rather than passive and accords far greater importance to prevention than to repair. Its key components include analysis and modification of social systems, including engineering environments and man-environment combinations, that maximize adaptation. Its person-oriented prongs stress such approaches as early childhood intervention, crisis intervention, and consultation which vastly extend outreach and promise for nearly geometric pay-off increments from finite sources.

A recent book by Rosenblum (1971) argues strongly that prevention and positive development should be the central focus of community psychology and mental health. And Atthowe (1973) has recently considered this question specifically with regard to mental health type problems.

Assumption VI

Human problems usually come in clusters and are the consequence of multiple interacting conditions. Therefore, effective programmatic approaches to human concerns must recognize and deal with the fact that problems usually occur in clusters and are multiply determined. For example, a deeply depressed woman is referred to a mental health clinic by her minister for help in alleviating her depression. Why is she depressed? Her husband is sick, her son is on drugs, she is caring for an aging father, and she has no friends or social contacts outside the confines of her home. Some crisis intervention may momentarily alleviate her depression, but unless the stresses of her life somehow or another change, she is probably going to continue to be at least periodically depressed. Effective intervention to help her will probably require dealing both with her personal reactions and some attributes of her day-to-day environment. Because problems come in clusters and are multiply determined, multiple and correlated interventions should probably be the rule rather than the exception in any approach adopted.

Assumption VII

The effective correlation of multiple approaches will require a correlated rather than a categorical human services delivery system. One increasingly strong conviction about the current situation with regard to mental health and other human services is that the present fragmented and categorical approach is excessively costly and insufficiently effective. The present categorical approach tears people and their lives into pieces, and assigns jurisdiction for each piece to some separate administrative structure and professional guild. Do you have a mentally ill child? We have an office over here for that. Does the child have some motor coordination and development difficulties? There's a physical therapy service in the next block. Is her dad out of a job so he doesn't have money to help the child? Across town we have an office with income maintenance programs.

Down the street we also have a vocational rehabilitation office to help him prepare for a new job, and in the other block we have an employment bureau to help him find a job. Is the mental illness correlated with a biological vulnerability which was partly the result of ignorance in the use of diet or drugs by the mother when pregnant? We have a nutrition service over there, and a drug abuse program over here to help the mother avoid such difficulties in the future. Is this mentally ill girl in trouble with the law? Then, she is a juvenile delinquent and we have a separate service for delinquents.

The structure of our service delivery system is not designed to fit the manner in which we lead our lives, develop difficulties, or correct them. We have applied the economic principle of the division of labor to develop specialists for our eyes, ears, bodies, work skills, stomachs, minds and hearts; and then we have organized our services as if those parts can be treated separately from one another — *and they can't.* Our present categorical approach does not make the best personal, social or economic sense for the future. For example, child development and day-care programs have been increasing. Is that a welfare program? Partly. Mental health? Yes. Physical health? Yes. A supporting service for parental employment? Maybe. In other words, this one type of program would appear to fit several typical categories.

The magic word to cover our present fragmentation of services is "referral". But the fact is that referral from one specialized agency to another doesn't work well, is often not accomplished, and the client too frequently gets caught in the crack between agencies. If coordination of services takes place now, it is often the client who has to do the coordinating, because specialized services usually attend only to the problems in their domain of competence, and tend to protect their "turf" in relationship to other services. Each service may work to try to deal with its piece of the cluster of the problems and contributing conditions, but unless close correlation occurs, they may in fact duplicate or even interfere with and disrupt the efforts of one another.

The concept of mental health centers was, in my opinion, a serious step toward correlation of services (e.g., Glasscote, Sanders, Forstenzer, and Foley, 1964). It has experienced restricted success, however. It still involves a categorical approach and if it tries to broaden itself into the very broad concept of mental health which some propose, it immediately runs into other categorically defined services with their own funding and turf patterns. Some kind of more comprehensive united services approach must be evolved if the assumptions proposed in this paper are adopted. Experiments in this direction are beginning to occur all over the United States.

One potential benefit of adopting a united services approach could be the gradual abandonment of the kinds of labels now placed on people as the result of our categorically defined delivery system — labels which we all know can themselves produce destructive consequences (to which a recent national Vice-Presidential nominee can attest). The present approach is socially destructive because it tends to dehumanize people by identifying them as a problem category rather than as persons in a social context.

Assumption VIII

The kinds of problems to which the labels "mental health" or "human services" are directed affect people in all socio-economic levels, therefore, the service delivery arrangements should be designed to serve all segments of the population. We should not have one pattern of services for the poor (welfare services), another for the rich (private fee for service), and another for the middle class (community mental health clinics). A delivery system designed to ignore socio-economic class differences and to avoid labeling people with what have come to be socially demeaning diagnostic labels will, in its form and function, promote some of the most fundamental social values of our culture, while at the same time help people deal with their personal concerns.

The nature of the service provided and the system through which it's provided should be a separate issue from the manner of financing the services for each individual. Some may pay for all service, while others may be assisted entirely with a public subsidy, but the service itself should be available to all without distinction as to the source of financing the service.

Assumption IX

The crucial place for this conceptual and methodological correlation to occur is at the local community level; that is, at the front line point of delivery. What is needed is something analogous to a public school system where experts in English, chemistry, mathematics, biology, history, psychology, and the like are organized to provide growth experiences to young people as they progress through different stages of development. Can we imagine an analogous community approach to bio-psycho-social development? It should be possible to design correlated approaches at the local level even if the super structure at state and federal levels does not have the same degree of coordination. All that is necessary is sufficient flexibility in the use of funding streams when they arrive at the local level to permit correlated approaches. General and special revenue sharing, if adopted and expanded, may provide such flexibility.

Assumption X

One of the greatest present weaknesses in developing sound mental health in human services programs is in defining clearly the results sought and in evaluating effectively the results achieved. We have available a great many ideas about how human behavior works, the forces which affect it, and a sizable variety of intervention strategies. What we do not have is sophisticated conceptual frameworks and technologies by which evaluation of the impact of various kinds of social intervention can be made in a convincing way. We generally count activity indicators (e.g., how many beds are full, how many people visited the clinic, how many types of services are provided) rather than outcome indices (e.g., in what ways have clients increased their daily living effectiveness, how has work productivity increased, how has subjective experience improved, what decline in specific types of problems has occurred in a given population of people). Until evaluation strategies focused on outcome variables are improved and used, programs will continue to be shaped by factors such as who has the loudest and most forceful political voice, rather than knowledge of the relative effectiveness of different approaches for different purposes. A budding national thrust on this issue (H.E.W., 1969) has not flowered, but as accountability becomes a basic social policy watchword, evaluation will become its handmaiden.

Assumption XI

The nomenclature adopted for the next evolution of mental health and human services should emphasize the positive development orientation of our culture and minimize the possible negative consequences of stereotyping that are a result of labeling with terms having a derrogatory social connotation. Moreover, the labels adopted should not be ones closely identified with particular disciplines or professions, because such nomenclature will serve as a barrier to producing a cooperative effort among the necessary multiple disciplines and professions.

Labels such as human resources services, or human development services illustrate terms with positive connotations and without particular discipline or professional ownership at this point. Others are possible, of course. But, development of a new coordination and synthesis will be greatly facilitated, in my judgement, if a new nomenclature can be brought into existence, acceptable to the multiple forces which will have to come together to accomplish this purpose. If the term "mental health" is to continue to be used, I would suggest that it be used for a narrower category of concerns than that toward which it has been tending, probably as a result of its roots in "mental illness".

SPECIFIC TARGETS FOR ACTION

In closing, I would suggest four targets on which our actions should be focused during the next few years.

Development and Evaluation of Intervention Strategies

Regardless of the conceptual frameworks and delivery systems which emerge in the future, scientifically-validated intervention strategies, to achieve clearly identifiable objectives with evaluation strategies for demonstrating the results achieved in practice, must be more fully developed. We should encourage, support, and participate in efforts to accomplish this objective.

Evolution of Integrative Theories of Human Functioning

New conceptual frameworks, which more adequately represent the person as a complex bio-psycho-social system in continual transaction with a variety of environments, must be developed. Such conceptual frameworks will have to be multi-disciplinary in origin. Therefore, we should encourage, support, and participate in efforts to evolve such integrative conceptual frameworks and scholarly contexts.

Evolution of New Delivery Systems

If people are to be treated as persons in a social environment rather than as a collection of parts or as categories of symptoms, it will be necessary to evolve new patterns for the delivery of services which erase the existing barriers among present separate, categorically defined, and guild-controlled service systems. Moreover, if a serious commitment to prevention and positive development is to be implemented new delivery systems will have to emerge which support and facilitate such efforts. Therefore, we should adopt professional attitudes which eschew the defensive posture of protecting our personal and turf identities and securities, and which focus on cultivating new, more effective arrangements. We should support, and lead in the effort toward a new, more effective synthesis, even if it means giving up the identity, terminology, and political leverage of "mental health".

Development of the Next Generation of Professionals

The approach toward which this paper points (and toward which I believe our society is moving), will require scholarly and professional leadership with somewhat different (and conceptually and methodologically richer) backgrounds than our generation obtained. We must provide leadership for, support, and participate in efforts to evolve new educational patterns for this next generation. This means, for example, that training programs must not be narrowly accredited, or certification or licensing procedures defined as guild defenses.

Unless we move forcefully toward these targets, with the highest degree of scientific rigor, professional sophistication, and personal or professional selflessness, (which change often demands), we will fall short of our social ideals and our own creative potential. I suggest we recognize the term "mental health" as a label for an important social goal and cultural value (Ginsberg, 1955), and develop broader, more appropriate terminology for the conceptual, programmatic, and professional activities associated with remedial, preventive, and developmental human services.

REFERENCES

Atthowe, John M., Jr. "Behavior innovation and persistence." *American Psychologist*, Vol. 28, No. 1, January, 1973.

APA Monitor, Vol. 4, No. 4. Washington, D.C.: American Psychological Association, April 1973.

Cowen, Emory L. "Social and community interventions." In *Annual Review of Psychology*, Vol. 24., 1973, pp. 423-460. Palo Alto, Calif: Annual Reviews, Inc.

Encyclopedia Britannica. "Mental Health." Vol. 15, 1966, p. 171.

Ford, Donald H. and Urban, Hugh B. "Psychotherapy." *Annual Review of Psychology*, Vol. 18. 1967, Palo Alto, Calif: Annual Reviews, Inc.

Ginsberg, Sol W. "The mental health movement: Its theoretical assumptions." In Kotinsky, Ruth, and Witmer (Eds.), *Community Programs for Mental Health.* Cambridge, Massachusetts: Harvard University Press, 1955, 1-27.

Glasscote, Raymond, Sanders, David, Forstenzer, H. M., & Foley, A. R. *The Community Mental Health Center: An Analysis of Existing Models.* Washington, D.C., American Psychiatric Association, 1964.

Golann, S. E. *Coordinate Index Reference Guide to Community Mental Health.* New York: Behavioral Publications, 1969.

Guerney, B., Guerney, L., & Andronico, M. *Filial Therapy: A Case Illustration in Psychotherapeutic Agents: New Roles for Non-Professionals, Parents, and Teachers.* B. Guerney (Ed.), New York: Holt, Rinehart, & Winston, Inc., 1969.

Hardin, Garrett. "The tragedy of commons." *Science*, AAAS, Vol, 162, No. 3859, December 13, 1968, 1247-1248.

National Institute of Mental Health, *Mental Health Program Reports No. 2; Public Health Service Publication, No. 1743.* U.S. Department of Health, Education, and Welfare, U.S. Government Printing Office, Washington, D.C., 1968.

Pike, Kenneth. "Language in relation to a unified theory of a structure of behavior." *The Hague*, Mouton and Co., 1967.

Rosenblum, Gershen (Ed.), *Issues in Community Psychology and Preventive Mental Health.* New York: Behavioral Publications, 1971.

U.S. Department of Health, Education, and Welfare, *Toward a Social Report.* U.S. Government Printing Office, Washington, D.C., 1969.

Wechsler, I. S., "The legend of the prevention of mental disease." *Journal of the American Medical Association*, 1930, Vol. XCV, 24.

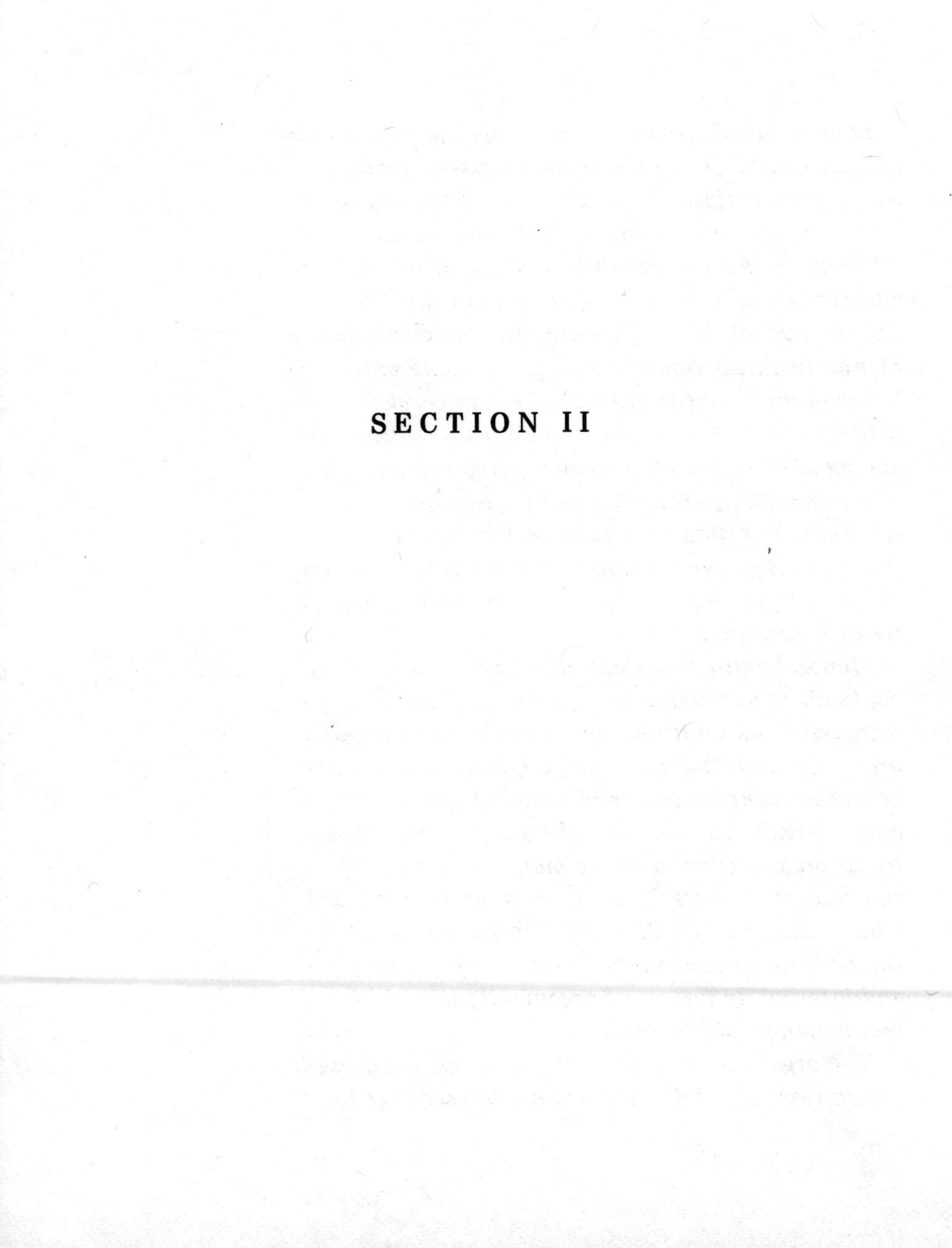

SECTION II

Perspectives On Delivery Systems

Mental health organizations, such as community mental health centers, can be discussed from many perspectives. As in the case of the six blind men touching the elephant, each perspective is likely to have its constraints, and result in a limited and particular definition of reality. In the chapters which follow, we are interested in viewing mental health organizations, and to some extent larger human service organizations, as service delivery systems; i.e., as complex organizations having purposes related to enhancing the quality of human adaptation. Our perspective on these systems will be primarily one of analyzing their functioning as organizations rather than, for example, reviewing the quality of services provided by these organizations.

Mental health organizations might first be conceptualized as involving a structured flow of resources which ultimately results in certain end products, with the latter ranging from community crisis intervention to a mental-health-center-office coffee break. That is, both the services provided by an organization, and the maintenance of the organization itself, are important ends toward which resources are directed. For purposes of the present discussion, however, the problems and internal dynamics of organizational maintenance will be omitted.

The organizational flow of resources, resulting in the conversion of dollars into human services in a mental health organization, involves at least three major levels of system functioning; each level has its own unique problems and dynamics. Level one involves the initial exchange of resources between the resource provider and the service-providing organization. The individual identities of the resource providers might range from a wealthy philanthropist, to a county court, to the Department of Health, Education, and Welfare. However, to the mental health organization, the collective identity of resource providers is more often than not their own mental health center board, or the governance structures of related human service organizations. The nature of the relationship, both contractually and informally, between an organization and its governing board is of critical importance in establishing the parameters and contingencies which surround that organization's staff behavior and program development.

A second level of analysis is that of the sometimes facilitative, sometimes problematic, relationship between service providers themselves. That is, how do service providers structurally and programmatically organize their efforts, particularly in terms of the myriad of inter-organizational problems that must be confronted and dealt with in human service organizations? It is at this level of analysis that, it is often alleged, unnecessary duplication of services occurs. Or occasionally, inter-organizational arrangements are negotiated enabling two human service delivery systems to provide quality services at a cost less than it would be if each organization separately provided those services. As an exercise, the reader might try to think of a few examples of the latter. Allow yourself considerable time.

A third level of analysis involves the primary purpose for which the organization was designed; i.e., the organization's principal services. At this level of analysis there is the basic interchange between the provider and the consumer of the services. Of course, in addition to the nature of this interchange and the characteristics of both clients and service programs, there is the problem of the extent to which a system has been designed so that the resources remaining for actual services, after having gone through the conversion processes of the two preceding levels, are adequate. Too often, unfortunately, the "trickle down" flow of resources ultimately results in meager, understaffed programs.

Although we have no hard data on this subject, a review of the percentage of total organizational resources (i.e., their total budget) actually taking the form of client services in mental health organizations known to the authors, ranges from 35 to 50 percent. However, these figures assume that mental health treatment program staff actually do spend full time in treatment programs. Very often this is not the case. Perhaps a more accurate and realistic estimate of the "trickled resources" would be around 10 to 25 percent.

The chapters which follow in this section are linked to the preceding levels of analysis in the organizational conversion of resources to services. Reversing the earlier order of discussion, and beginning with the number-one priority of the client-organization relationship, Spaner discusses "New Directions in Community Mental Health Center Programs", and Baker examines the extent to which community mental health centers have, in fact, developed continuity of services in his chapter, "Are Community Mental Health Centers Organizing for Continuity and Efficiency?" The second level of analysis, involving relationships among service-providing organizations, is given attention in Baker and Broskowski's, "The Search for Integrality: New Organizational Forms for Human Services". Finally, the third level of analysis, that of the relationships between service and resource providers, is thoughtfully probed in Demone's chapter on "Influencing the Policy Decision".

Fred E. Spaner

Prior to 1963, and the enactment of the Community Mental Health Center Act, the treatment of people with mental health problems was largely institution-oriented. On a nationwide scale, similar to those afflicted with tuberculosis and leprosy, the person deemed by his community to be "mentally ill" was removed from his community for treatment.

It might be argued, as the Nader group in their report on community health centers has done, that the centers have fallen far short of the "bold new approach" to community programs promised by the legislation. However, what such an argument tends to overlook is the remarkable fact that in a relatively short period of time, probably between 1963 and 1968, an historical pattern of the treatment of deviance was effectively reversed. Both programmatically and conceptually, mental health delivery systems ceased to serve as routine community export agents.

In the following paper Spaner discusses the programmatic aspects of the Community Mental Health Centers Act; its nature and intent, and some of the more innovative recent developments which have come about because of this legislation. While Spaner's paper has a contemporary program focus, it might be viewed as a documentation or a current status report on the reversal of a 200-year pattern of deviance-management in America.

CHAPTER 4

New Directions In Community Mental Health Center Programs

INTRODUCTION

The Community Mental Health Centers Program (CMHC) is essentially a conceptualization of objectives for the delivery of mental health services, and operations for their implementation. It represents one of the first attempts at a fully coordinated system of care for a total population in a defined geographic area. In this paper I will attempt to discuss the general concepts and objectives of this program, their rationale and general indications of trends, based on visitations to federally-funded community mental health centers, the National Institute of Mental Health (NIMH) site-visit reports, and annual inventory data provided to the NIMH from the CMHC.

The Community Mental Health Centers Program, launched with the enactment of the Community Mental Health Centers Act of 1963 (Title II, 1963), has grown from a handful of centers first awarded grants in 1966, to 494 federally funded community mental health centers as of January 1, 1973. Of this number, 340 are currently operational and the rest are expected to be operational as soon as the facility for which they received a construction grant is completed. When all of these centers are operational, they will provide mental health services to approximately one-third of the Nation. These centers are located in all fifty states, the District of Columbia, Puerto Rico, and Guam. The first group of centers to receive staffing grants will be coming to the end of its eight years of federal support in fiscal year 1974.

It has only been seven years since the first implementation of the concept of a community mental health center, yet these concepts have become an integral part of our mental health language. The general context in which such concepts are applied can best be understood by understanding how a CMHC is organized and initiates operation.

A center receives a federal construction and/or staffing grant, usually, as a result of two or more existing community mental health service agencies joining together in developing an acceptable application. Hence, a community mental health center evolves from already existing patterns of mental health service delivery (Ozarin, Feldman, & Spaner, 1971). Few community mental health centers appreciate the full nature of commitment, when they accept their federal grant, to be responsible for providing a comprehensive range of mental health services, tied together in a manner to assure continuity of care, available, accessible, and responsive to the mental health needs of all persons residing in their catchment area.

The catchment area concept is considered by NIMH as the cornerstone of the federally-supported community mental health center program. Arthur D. Little, Inc. (1972), in an interim report of a study of the catchment area concept, concludes that:

> Centers tend to experience a natural evolution, . . . as they 'live into' the catchment concept, that engages the following sequence of themes:
> — geography and population phase;
> — accessibility and community connections phase;
> — accountability and service delivery phase.

Community mental health centers experience a natural evolution not only as they "live into" the catchment concept but as they "live into" each of the process objectives which are basic to the federally-supported Community Mental Health Centers Program. These process objectives are: comprehensiveness of services, continuity of care, accessibility and equity of services, community involvement in the planning decisions, and accountability and responsiveness of services to the mental health needs of the catchment area.

COMPREHENSIVENESS

The comprehensiveness range of mental health services has been formulated in regulatory language as five essential services: consultation and education, emergency, outpatient, partial hospitalization, and inpatient services. The then-existing service designations, when the centers' legislation was enacted, were used to define comprehensiveness. However, each of these essential services should not be considered as a discrete service, but rather as one of numerous bench marks along a continuum of services from prevention, to an immediate response to a mental health crisis, to an on-going arrangement for treatment and support, to more intensive care which is less than 24-hour care, and finally, should the condition warrant it, 24-hour care for as short a period as possible.

These bench marks have received different degrees of emphasis by different centers, depending on the treatment philosophy of staff and needs of the community. Each of these designated services changes as a CMHC develops.

Consultation and Education

This service has as its mission the reduction of direct services by the CMHC by using its staff to bolster the ability of community caretakers to handle the problems of people, which, if not recognized or responded to at an early stage could become serious enough to require direct service. Consultation seems to evolve from an initial case-orientation approach to a more readily generalized application of system consultation. Education often starts as information related to case-finding and orientation about the center's role, functions, and capabilities. As the center matures this may change to informing, and even instructing, the public about forces which may adversely affect their mental health; and in some instances, to helping citizens to understand how community stresses and conflicts can be reduced. Therefore, consultation and education seem to shift from a focus on individual identification of problem areas and attention to potentially vulnerable groups, to concern with the general quality of life in the community.

Consultation and education services change appreciably as a center tries to meet its catchment area's needs. Currently, NIMH data show that the most frequent recipients of CMHC consultation and education services are school districts (CMHC Inventory, 1970). More than 90 percent of federally-supported CMHC report consultative programs with school systems in their catchment area. Among other agencies with which CMHC staff consult, are courts, police departments, welfare agencies, housing projects, health agencies, and community planning organizations.

Emergency Services

These services are required on a 24-hour-a-day, seven-day-a-week basis, and may evolve from what is only available in the general hospital emergency room for all medical emergencies, to a mental health intervention capability which reaches out and is responsive to individuals or groups experiencing a mental health crisis anywhere in the catchment area. Some steps in the evolution of a CMHC emergency service are: first, the assignment of center staff to be on call to the general hospital emergency room; second, orientation and training of emergency-room personnel in how to respond to mental health emergencies; third, the establishment of an emergency telephone answering service for the catchment area; fourth, the assignment of mental health staff to the hospital emergency room; fifth, the development of a plan for emergency outreach to all parts of the catchment area; and sixth, the establishment of a response-capability for transporting a mental health emergency.

Making trained mental health staff available for an emergency in some sparsely populated, large geographic catchment areas may require the use of airplane services. Some urban centers have used two-way-radio-equipped mobile vans. The degree of sophistication of a center's emergency service depends on the kinds of emergencies in the catchment area, and the resources of the center.

The kind of treatment offered to the person in crisis also undergoes change over time. Most emergency services initially rely heavily on medication to tranquilize or sedate the person. It takes experience and a lessening of staff anxiety to permit the person in crisis to experience it, and use his own personal resources and controls to overcome the difficulty.

A fully effective emergency service should provide 24-hour, seven-day-a-week:

1) walk-in service
2) telephone answering service, and
3) outreach potential

Some emergency services are also developing a 24-to-72-hour bed capability which enables a person in crisis to resolve the problem, without having to be admitted to an inpatient unit. A large western metropolitan center found that a small number of 24-to-72-hour crisis-intervention beds reduce inpatient bed utilization by 80 percent.

The most recent direction for emergency mental health services is the development of community emergency plans. During the last year's Hurricane Agnes, thousands of families were left homeless in the upper Susquehanna River Valley. Two community mental health centers in Luzerne-Wyoming County recognized the mental health implications of this natural disaster and participated in a broad crisis intervention effort. These CMHC took responsibility for recruiting and training a cadre of 50 human-resources aides to canvass relocation camps, emergency trailer villages, and homes, in order to provide information, referrals to service resources, brief counseling, and to, generally, evidence a warm, humanitarian concern for their neighbors. Such community activities are an outgrowth of the emergency services requirement, as well as the consultation and education requirements, the basic catchment area concept, and the need for centers to be responsive to the community's needs.

Outpatient Services

Outpatient services are probably the most stable and best understood programs of CMHC. However, psychotherapy, even group psychotherapy, is impossible for all who may seem to require this service, when a center must serve all residents of its catchment area. Many centers have resolved this by having vast numbers of their clientele on medication so that outpatient services may, in many instances, become a monthly, ten-minute, medication/supervision session. Such a center then reserves, for a small number of selected, young, verbal adults, the bulk of the outpatient staff hours in individual and group psychotherapy.

As centers evolve, many become more venturesome and attempt to reach larger numbers of their clientele on an outpatient basis through other types of psychotherapeutic relationships. This may include the use of large numbers of mental health workers and volunteers as therapists, the development of leaderless groups, and the encouragement of caretaker agencies in the community to take responsibility for conducting groups.

Other new directions include behavior modification programs, psychotherapy which emphasizes the here-and-now, and continuous groups. An example of the latter is a program in a metropolitan center which has a large, old, three-story mansion as the CMHC headquarters. On the second floor is a large parlor, which is used as the meeting place for a continuous group. Staff and patients come and go at will. They can talk or remain silent. A member of the staff is always available. Any person from the catchment area can enter. This room, in some instances, serves as the waiting room for clients who have other business with the center. In other instances, a person, new to the center, might enter the group and have his intake interview within the group, if he agrees to it. This center is always open, and staff are assigned duty coverage around the clock. There aren't many centers which maintain that kind of open schedule.

Because of the long-established patterns in outpatient services, new directions for ongoing care for persons in need of mental health services have not been subject to the same degree of experimentation that has occurred in the less-well-established portions of the comprehensive range of mental health services.

Partial Hospitalization

This service consists of care which is more intensive than outpatients' care, and less intensive than inpatient care. NIMH has required that, at a minimum, each federally-supported CMHC have a day-care program. These programs are quite varied. They may be activities largely reminiscent of occupational therapy on an inpatient unit, or contract work for pay in a semi-sheltered workshop. Partial hospitalization is a misnomer since it does not have to occur in a hospital. Rather, it refers to mental health services which bridge the gap between services provided on an-hour-a-week basis, and services provided on a 24-hour-care basis. Partial hospitalization programs may be day-care, night-care, or weekend-care programs. They are intended mainly as alternatives to 24-hour, inpatient, care.

The more real-life-oriented these programs are, the more they become a part of, and are supported by, the community. In Seattle, a mental health center's day-care program operates a gasoline service station in addition to its other activities. Thus, in conjunction with the center staff, day-care patients work at their own pace in real jobs and in an understanding atmosphere. They are able to test their ability to work and relate to others in this real life situation. In a center in Philadelphia, the night-care program starts for its clientele after 5:00 p.m., and terminates in the morning after breakfast. The evenings are spent in learning how to relate to other people in a non-structured, non-work-like-atmosphere. Most partial hospitalization programs are time-limited, and clients needing care after the allotted time, are usually continued on an outpatient basis or referred to sheltered workshops, halfway houses, or family-care facilities.

The principle new direction for partial hospitalization programs is care of the seriously-disturbed individuals. There now seem to be several demonstrations that severely-disturbed individuals can profit from these programs without having to be admitted to inpatient units. Some centers are attempting to achieve a zero inpatient length of stay by the use of partial hospitalization programs.

Inpatient Services

Inpatient services usually provide 24-hour care in a hospital setting. Most CMHC have learned that they overestimated their need for inpatient beds. A center usually evolves from a cautious posture which initially assumes that whenever in doubt it is wise to hospitalize, to a position of lessened staff anxiety which uses inpatient admissions only as a last resort. Inpatient services in CMHC are showing a dramatic reduction in length of stay and are reducing the use of state hospital beds by catchment area residents (1970). Some other new directions involve the use of other than hospital beds for 24-hour care. These include the use of motels, half-way houses, nursing homes, and supervised group living.

The therapeutic milieu on inpatient services is also evolving. Many CMHC use a portion of their partial hospitalization as an adjunct to their inpatient unit. Thus, the inpatient unit is only used for sleeping, and morning and evening meals, while the partial hospitalization program serves to provide daily activities. This helps to coordinate these programs and broaden the scope of the two staffs. One CMHC has as its primary focus a day-care program which, in addition to occupational-like activities, has a larger parlor area with ten to twelve couches. Each of these couches converts into a bed. If individuals in the day program need to remain overnight, they sleep on these converted day couches. This then becomes the inpatient service for the center. The same area and staff also serve as the emergency service. Thus, anyone may walk in on a 24-hour, 7-day-a-week basis and have immediately available, within the same space and staff, the full range of direct services.

The new directions and innovations in providing comprehensive services are not without potential problems, the principle one being how to support services for which there are no existing reimbursement programs. Current third-party payment criteria are bed and visit oriented. If such services continue to be the only ones to be reimbursed, then according to the adage that "treatment follows the buck", the present trend toward finding alternatives to 24-hour care could very well be reversed.

The experience of the CMHC with comprehensiveness of service and some of the directions mentioned are at too early a stage to determine whether they will have an important influence on changing service reimbursement criteria. There is also a critical lack of evidence as to whether or not some of these service directions can show greater cost-effectiveness or greater social benefit than more traditional service patterns.

Comprehensiveness of service as a process objective is all-encompassing. The five essential services which I refer to as bench mark designations, do not specify "what" is to be provided to the client as much as "where and when" it will occur. Specific mental health services in terms of treatment may be similar in any of these service components.

CONTINUITY OF CARE

Since community mental health centers are service networks consisting of two or more agencies, coordination of their collective efforts to provide a unified system of care presents problems. The development of a non-fragmented system of care has attempted to take into account how the information, staff, and individualized treatment program for the client will be maintained, regardless of his movement from one service component to another. Centers have attempted a variety of solutions to this problem. These solutions differ according to the complexity of the center, the number of previously completely autonomous agencies providing the services, the interpersonal relationships of staff, and the degree of central leadership.

Record systems at CMHC may change over time from separate record systems for each component part to a unified single-record system for all parts of the center. A client may have a different staff member assigned for each service, or one staff member may be responsible for him regardless of the service. A separate and possibly different treatment plan may be established for the client on each service, or one treatment plan developed which is applicable across services. There may be a separate intake procedure for each service, or one intake into the system provides the client access to any service which may be needed. Some centers assign an advocate, or treatment coordinator, to each client on intake, who assures that continuity of care will be maintained.

The evolution of continuity of care differs greatly from one center to another. However, the experiences of each center leads it to recognize the importance of service coordination and promotes the development of mechanisms which will assure the client that his or her mental health experiences are known to each staff person responsible for the provision of care; also, that such care is part of an individualized plan which is known to, and considered by, whomever provides the care.

ACCESSIBILITY AND EQUITY

A community mental health center receiving federal support must have its services accessible to all persons residing in the catchment area. This means not only geographic accessibility, but also temporal, psychological, cultural, and fiscal accessibility.

Most centers find that geographic accessibility can only be achieved by developing an outreach program. Depending on the geography of the catchment area, this may require the center to establish satellites, outposts, or other sub-units of the parent center. Physical accessibility must also consider making mental health services available to persons who are unable to come to the center. The average number of home visits by center staffs increased by over 30 percent from 1970 to 1971 (1972). This shows an evolution for the national program as a whole.

The times when services are available is an integral part of accessibility. Emergency and inpatient services are the only ones which, either by regulations or definition, must be available on a 24-hour basis. The extent to which outpatient and partial hospitalization services are available at other than daytime hours, five days a week, is an index of temporal accessibility. Thirty-one percent of the centers have partial hospitalization programs overnight as well as during the day, and another nine percent have such programs in the evenings as well as days. Forty-six percent of all centers have day and evening hours for their outpatient program (Statistical Note).

Psychological and cultural accessibility relates to services being offered under conditions which do not inhibit traditionally underserved groups in the catchment area from seeking services. A center becomes sensitive to these factors as it continues to operate and keep track of the nature and composition of its clientele. If any significant group residing in the catchment area is under-represented among its services' recipients, then it merits examining whether or not the center is psychologically and culturally accessible to them. Many centers with significant numbers of non-English-speaking persons in the catchment area have recruited bilingual staff members at all center levels to permit them to work effectively with this clientele.

Fiscal accessibility is required in the federal CMHC regulations which state that a reasonable volume of services must be provided to residents of the catchment area, regardless of their ability to pay. In 1971, 63 percent of the additions to CMHC roles had incomes of under $5,000 a year. Only 19 percent of the U.S. population have such low income levels (1970).

The accessibility of services can be inferred by the number of persons referred by self, family, or friends, and the number of additions of persons without any previous mental health services.

In 1971, 37 percent of all CMHC referrals were by self, family and friends. This is a five percent increase over 1968. The total additions to centers in 1971 of persons with no previous mental health services was 51 percent. This is an increase of 14 percent over 1968 (1970).

The passage of the 1970 amendments to the CMHC Act provided an increased rate of funding for centers located in poverty designated catchment areas. Thirty-six percent of all poverty catchment areas have federally funded CMHC; whereas, only 14 percent of non-poverty catchment areas are covered by CMHC.

COMMUNITY INVOLVEMENT

Community mental health centers are aware that regulations require that they involve the community in the planning and operation of the center. Many find this difficult to implement. Mental health agencies have usually perceived community involvement as using volunteers, or having community leaders on their boards, or joining with community groups to obtain mental health program support from legislators. All of these important activities may serve as initial kinds of community involvement, but a CMHC must go beyond that.

As a CMHC attempts to reach all of the people in its catchment area it finds that community involvement must include many new constituencies, some with different agendas than those of the CMHC staff and administration. The largest group of CMHC grantees are general hospitals. They are usually governed by elitist boards who supervise the business and fiscal management, leaving service delivery (the product) to the professionals. The evolving pattern of community involvement brings more of the CMHC consumers into the picture, and their concern is more often directed at the services, (the product), than at the business part of the operation.

In general, CMHC wish to have input from the broad community, yet the professional staff and governing boards do not wish to commit themselves to having to follow community dictates. One of the mechanisms for achieving a compromise has been the formation of community advisory boards. These are either advisory to the director of the CMHC or to its governing board. The members of the advisory boards are chosen by community groups. They may be representatives of agencies, or have been selected from neighborhoods or other sub-divisions of the catchment areas.

The variety of patterns which have been developed are indicative of the difficulty of this problem. Advisory board members learn about their community's needs and their center's capability to meet these needs. As this process continues to evolve, some advisory boards take on more and more the role of advocate. Also, as greater numbers of lay people become knowledgeable about mental health services delivery and their relevance to their problems, they attempt to take a more active role in formulating the CMHC policy. This has produced conflicts in community mental health centers. In some instances it is more than the CMHC can handle. Some grantees have requested transfer of federal grants to community groups. This frees the hospital or university of the responsibility for operating the center and relieves them of the task of dealing with the community in an adversary role.

The community board can govern the center and purchase services from the hospital (the previous grantee). In other instances, hospital boards have changed their by-laws so as to include a greater number of catchment area residents. Another way in which community involvement has developed is through state legislation which establishes mental health boards throughout the state. Depending on the size of the jurisdictions of these mental health boards and the extent to which their regions are coterminus with catchment areas, they may be able to substitute for CMHC community governing boards.

Recent NIMH studies of citizen involvement confirm that this process objective is proceeding at a slower pace than others (Community Change, Inc., and Public Sector, Inc., 1972). However, a sample of site visits to CMHC show that from 1968 to 1972 (1972), two-thirds of the centers visited had increased their citizen involvement. Overall, in June, 1972 only 10 percent of the centers showed no citizen participation, 55 percent showed some, and 35 percent showed considerable involvement either on advisory or governing boards.

The direction is toward greater community involvement, but this has been largely a result of NIMH's requirement. Whether this direction would be maintained in the absence of NIMH's surveillance is problematic. Hopefully, CMHC will recognize its importance to their own viability regardless of grant-related requirements.

ACCOUNTABILITY AND RESPONSIVENESS

All of the objectives discussed contribute to the overriding objective of the Community Mental Health Centers Program — responsiveness to the mental health needs of persons residing in the catchment area and accountability of the CMHC to the community it is to serve.

Comprehensiveness of service, continuity of care, accessibility and equity in the delivery of service, are all aspects of responsiveness. Community involvement is one mechanism for both holding the CMHC accountable and permitting community input so that the program may be optimally responsive.

The degree to which a CMHC is accountable and responsive evolves in a similar manner to that of the other process objectives. However, this process is more than the summation of the others. As a separate objective, it requires a system which will enable the center staff, its administration, its governing body, and the public to know what services are being provided to whom, feedback as to their adequacy and relevance, and a way to use the feedback to modify and improve the program.

Most CMHC initiate this process when they obtain a wall map of their catchment area and mark it with color-coded pins to show how many persons from what part of the catchment area, with what characteristics, are being served in what manner. More sophisticated systems are available, but this initial type of approach can attempt to raise some responsiveness questions; such as, are there underserved areas or groups?

The NIMH contracted to have the 1970 census demographic data arranged on a catchment area basis (1972). This will be distributed to each state, as it is available, and the state in turn is to provide it to each

CMHC. This, it is hoped, will help the CMHC to assess their responsiveness.

Data and evaluation technology are one aspect of developing the capability for a CMHC to be accountable and responsive. The other is the center and the community jointly planning, setting priorities, reviewing use of resources, and determining objectives. How to most effectively identify and meet the mental health needs of persons residing in its catchment area, and provide such needed services within its fiscal resources, is the greatest challenge facing CMHC.

If a CMHC program has been evolving to achieve the process objectives discussed, then this problem should not be solely a challenge to the CMHC; it should also be a challenge for the community.

SUMMARY

From this brief overview the process objectives of the Community Mental Health Centers Program appear to be evolving in a manner consistent with the concepts initially formulated and funded. Currently available data is not sufficient to show the extent to which such objectives have been accomplished, or the outcome results, other than the reduction of the number of patients in state hospitals and the increased use of mental health services by more persons. An adequate system to assess the social and cost benefits of the CMHC program needs to be developed in order to determine whether its current directions can and should be continued.

REFERENCES

Arthur D. Little, Inc. "Evaluation of the Catchment Area Concept in the Community Mental Health Centers Program", Contract No. HSM-42-72-96, November 20, 1972.

Claritas Corp. "1970 Census Data Used to Indicate Areas with Different Potentials for Mental Health and Related Problems", Contract No. HSM-42-69-97, 1972.

Community Change, Inc., & Public Sector, Inc. "A Study of Consumer Participation in the Administrative Processes in Various Levels of HSMHA Service Projects", Contract No. HSM-110-71-135, June 20, 1972.

Data from Annual CHMC Inventory for 1970, Survey and Reports Section, Biometry Branch, Office of Program Planning and Evaluation, National Institute of Mental Health.

Ozarin, L. D., Feldman, S., & Spaner, F. E. "Experience with community mental health centers." *American Journal of Psychiatry*, 127:7, January, 1971, pp. 912-916.

Spaner, F. E. "Study of Forty-five CHMC Changes in Citizen Participation over Three-Year Period." Paper presented at the Annual Meeting of the American Psychological Association, September 1972.

Statistical Note, Survey and Reports Section, Biometry Branch, Office of Program Planning and Evaluation, National Institute of Mental Health, 1972.

Title II, Mental Retardation Facilities and Community Mental Health Centers Construction Act of 1963 (Public Law 88-164) as amended.

Frank Baker

In the preceding chapter Spaner posed continuity of care as a major goal of both community mental health legislation and most community mental health centers. The question remains, is it being achieved?

In the following chapter Baker[1] *suggests some interesting answers to the question of program continuity. In addition, he alludes to some important considerations in organizational design that are likely to be major determinants of whether or not community mental health centers do in fact achieve continuity of care in their programs.*

From the perspective of organizational design, Baker's description of the structural problems in providing continuity of care may have important implications for the design of Health Maintenance Organizations; organizations which will also have continuity of care as an important goal. It is likely that the structural development of HMO's will in many ways parallel that of community mental health centers. Consequently, we need to take a very careful look at the track record of community mental health centers in effecting continuity of care. The following chapter is a step in that direction, as well as a critical examination of an important problem for community mental health centers.

CHAPTER 5

Are Community Mental Health Centers Organizing For Continuity and Efficiency?

As is true for most of the goals which NIMH has specified for the Community Mental Health Centers Programs, the degree to which the Centers Program is succeeding in organizing for continuity and efficiency is very difficult to evaluate at this time. The difficulties include: 1) the definitional ambiguities in the conceptualization and operational definition of "continuity" and "efficiency"; 2) underdeveloped methods of measurement; and 3) sparseness of relevant research data and specific studies. This paper will briefly review definitions of the concepts of continuity of care and efficiency, related measurement problems, and the evidence from available research as to the extent to which the Centers Program is accomplishing these process goals.

CONTINUITY OF CARE

Definition

In one of the earliest and best discussions of the concept of care, Schwartz and Schwartz in 1964 noted that even at that time the idea (although not the general practice) of continuity of care of mental hospital patients was by no means new. They observed that changes in professionals' beliefs about mental illness and the role of the mental hospital were leading to greater interest in continuity of care, but they also observed that "it can only be achieved through appropriate organizational arrangements". They identified five elements in arrangements for continuity of care: 1) links between all the patient's helpers, 2) obtaining the patient's cooperation in entering aftercare, 3) providing for the transition between settings or types of care, 4) appropriate timing of contacts, and 5) relating aspects of the patient's previous experience to the further care planned for him. Schwartz and Schwartz (1964) noted that the continuity of treatment can imply continuation of a relationship with the same organization, in relationship with the same helper, the same help with a change in helpers, or a continuation of a program of help.

More recently Jepson (1970) suggested that program responsibility is more important than having one professional person maintain a continuing relationship with the patient; or, in other words, the ready acceptance of a patient between services and the free flow of information. Jepson concluded:

> No matter what phase of illness the patient may be in or what treatment modalities he may require, someone or some group, or some agency must maintain a concern and responsibility for him. This is at the same time a personal concern, a medical responsibility, and an administrative demand. When the total program with all its service elements operates under a unitary administration, this continuity of responsibility can be assured, and those all-too-frequent discontinuities of responsibility that occur when patients are transferred from one facility to another can be avoided.

For Jepson, the assurance of continuity of responsibility requires that each element of a program have the same inclusive admission policy that the overall system has and each service unit must accept any case sent to it by another unit.

Wilder, Levin and Zwerling (1970), recognizing that community mental health centers differ in important ways that affect continuity, have observed that "solutions to the problem of providing continuity of care will vary according to the needs of the patient, the mental health resources of population areas served, and the geographical location of facilities."

Bass and Windle (1973), commenting on the variety of mechanisms that community mental health centers used to promote continuity, describe results of an examination of a sample of 49 center applications for continuation of NIMH staffing grants which were received before July, 1969. The most frequent response (65 percent) of centers in describing how they maintain continuity of care between services was in terms of written and verbal communication mechanisms, while the next most frequent response (14 percent) dealt with the maintenance of a continuing relationship between a staff member and the client.

Glasscote, et al. (1969), in one of the first comprehensive studies of federally funded community mental health centers, suggested a dual approach to the use of the term continuity of care: 1) continuity of therapists in situations in which the patient is treated by the same therapist throughout his illness, and 2) continuity of responsibility in which a facility continues to accept responsibility for the patient and see to it that at any given time some element of service is responsible for him. Gray (1969) listed four major types of continuity: 1) continuation of a relationship (same as continuity of therapists); 2) continuity of services, defined as care continuing, although the caregiver, whether agency or therapist, changes; 3) continuity of program, characterized by a treatment plan jointly developed by those involved with the patient or client; and 4) relationship to an organization, meaning that one agency is responsible for and coordinates the treatment provided to individual clients or patients.

In summary, the concept of "continuity of care", although widely recognized as a goal of the community mental health center program, has lacked a single generally recognized operational definition.

Measurement

In one of the first published studies relevant to the quantitative measurement of continuity of mental health services Pugh and MacMahon (1967) examined discontinuity in 29 public and private Massachusetts hospitals. Their operational definition of discontinuity was the transfer of patients from one inpatient facility to another or readmission of an individual to a different mental hospital from the one to which he had been previously admitted. They used this measure to study the relation of discontinuity to diagnosis.

Recently, Bass and Windle (1972, 1973) have collaborated in several studies which have done much to advance the sophistication of measurement of continuity of care. They define continuity of care operationally as existing to the extent that "there are no obstacles to a client's either remaining in, or moving from, any of the center's direct treatment services in conformity with his therapeutic needs; and that administrative mechanisms relate past and present care by providing stable client-caretaker relationships, necessary written and verbal communications among staff members, and contact with clients who miss appointments, go on unauthorized leave, or otherwise appear to be dropping out of treatment prematurely" (Bass and Windle, 1972).

Their definition highlights four criteria of continuity:

1) client movement or the absence of it in appropriate response to treatment needs;
2) stability of the client-caretaker relationship;
3) both verbal and written communication among staff members;
4) efforts made to retrieve clients who appear to be dropping out of treatment prematurely.

Bass and Windle (1973) report a preliminary effort to measure continuity of care which they describe as a combination of Pugh and MacMahon's quantitative approach and Gray's broader coverage of junctures in the treatment program of a center. Bass and Windle obtained retrospective information on the direct service treatment history of 947 adult clients of a community mental health center during a one-month period on the basis of an examination of case records and the recollection of therapists. They were able to examine whether clients during a one-month observation period moved from or remained in a treatment service in response to treatment needs. However, the recollections of therapists and case records were found to be inadequate to provide information on other indices of continuity which they had generated from their attempt to develop an operational definition of continuity of care. They concluded that the inadequacies of data in records and the difficulty encountered in obtaining retrospective data on clients' movements, and on reasons for decisions made in relation to clients, required a prospective approach rather than a retrospective one.

Bass and Windle (1972) went on to develop a multiple-choice questionnaire with a prospective orientation. This questionnaire which is called the "Continuity of Care Inventory" is comprised of five sections and is intended to be used during a limited observation period to measure the direct essential services of a center for continuity of care. As designed and tested by Bass and Windle, this instrument requires a considerable long-term collection of data in a center, and thus it tends to have somewhat limited applicability unless the organization being examined is willing to commit sufficient resources to institutionalize such measurement over time. However, Bass and Windle have performed a useful service in beginning to operationally define a key concept which can be useful in judging the performance of a comprehensive helping unit service center.

Research Evidence on CMHC Achievement of Continuity of Care

The National Academy of Public Administration (NAPA), in a study supported by an NIMH contract (NAPA, 1971), attempted to analyze the organizational and administrative relationships of nine multi-agency community mental health centers and evaluate the impact of these arrangements on the continuity of care of patients treated by these centers. During their visits to the nine centers, the NAPA study team found that continuity of care was a familiar concept to the staff of the centers but that the phrase was subject to varying interpretations and the methods which were used to achieve this goal differed significantly from center to center. A little more than half (53 percent) of the total group of respondents indicated that they felt the concept was "extremely" or "very well" understood.

The NAPA report stresses the contextual importance of individual center operations to detailed aspects of continuity of care. They observe that the method of delivering services in centers is less determined by overall guidelines or procedures than it is by the types and numbers of professionals involved, kinds of agencies participating, and numbers of clients who must be treated in a particular center. The NAPA report also points out that the newness of the program and its emphasis on innovation as well as the underdeveloped character of its information systems and data basis also militate against any precise measurement of continuity across centers on a comparative basis.

The NAPA study found that the staff of some of the centers visited by the research team rejected continuity of therapists on the philosophical grounds that it followed the "medical model" in its emphasis on a patient-doctor relationship and tended thus to create a dependency relationship. The National Academy of Public Administration team concluded that continuity of care can be provided without continuity of therapists, while maintaining that treatment responsibility need not be provided by the same therapists throughout the course of a client's treatment.

They also stated that there should be a clear assignment of responsibility during all phases of treatment and that the number of individuals who are made accountable for an individual patient should be kept to a minimum. Thus they took account of the fact that it is very often the practice in multi-agency centers for a single staff member to maintain treatment responsibility. Within the member agencies comprising a multi-agency center, the NAPA study did recommend that continuity be maintained either by a single therapist or a team who would be responsible for a patient or client as long as he or she was within that particular agency. In general, the larger and more complex a center, the more difficult is a provision of continuity of care according to this study. The NAPA study group observes that: "A center with numerous agencies and specialized facilities can make services more accessible and comprehensive, but these advantages may be obtained at the cost of continuity unless formal linkages and channels of communication are established to provide adequate coordination" (NAPA, 1971).

The National Academy also found that complexity of the centers they studied seemed to be related to ratings by the staff of how well continuity of care was being implemented in the center in which the staff worked. The lowest self-ratings were received by the two centers which were the most complex of the nine listed by the Academy research team. In general, this study found that only a small minority (18 percent) rated their community mental health center as implementing the concept of continuity less than adequately. Forty percent of the 411 staff surveyed rated their centers as implementing continuity of care "extremely" or "very" well.

A study of relationships between federally funded centers and public mental hospitals conducted under contract to NIMH by Socio-Technical Systems Associates (1972), although it did not specifically focus on measuring continuity of care, is relevant to this discussion. The STSA study found that, in general, reported relationships between the 198 centers completing a questionnaire in 1971 and the public mental hospitals (PMH) serving their same catchment area were at a relatively minimal level. These relationships which were defined by relative presence or absence of joint PMH-CMHC programs, similarity in ratings of ultimate service goals, detailed information exchange, and center provision of inpatient care were found to be closest between centers and mental hospitals which were both public facilities, had formalized affiliation agreements, and were physically located comparatively near each other. The results of this study lead to an inference that organizational auspices, presence of formal agreements, and location in relation to other service organizations are important variables affecting continuity of care in terms of inter-organizational relations.

An NIMH contract study by the National Opinion Research Center (Orden and Stocking, 1971) examined relationships between nine community mental health centers and other care-giving agencies including an analysis of the effect of interagency relationship in continuity. They concluded: "continuity across agencies seems more related to the administrative ties among regional, state, county or city or other local government, and voluntary organizations." The NORC researchers found that continuity within centers was related to the number of affiliates or affiliate agreements and to the "administrative ability of the person or department bearing the responsibility to provide sufficient staff and funds for record keeping and follow-up care". The measure of continuity developed by the NORC staff consisted of ratings by center staff on a five-point scale of satisfaction ranging from "very satisfied" to "not at all satisfied". They then employed the relative ranking of the nine centers studied based on the average scores for each center. Although they attempted to get ratings from other care-giving agencies, they found that the staff at these other care-giving agencies often did not feel qualified to comment on continuity.

NORC also developed a measure of interagency network by ranking centers on four variables: 1) the proportion of staff time devoted to consultation and education, 2) the proportion of consultation and educational staff time directed to care-giving agencies, 3) the proportion of referrals of new clients to the center from other care-giving agencies, and 4) the proportion of clients referred at the termination of treatment to other agencies. The nine centers studied by the NORC group were rated on each of the four variables separately and ranked from 1 (highest interagency network) to 9 (lowest interagency network) on the basis of their combined ratings. Correlating the interagency network rank and continuity ranks they found a moderate negative relationship ($r = -.47$).

The NORC group expressed surprise at this negative correlation and suggested that it might be a function of the size of professional staff since three of the centers with the highest number of full-time staff rated themselves lowest in continuity suggesting that those centers which are highly professionalized tend to be more critical or have higher expectations for continuity. They offer an alternative explanation that a more complex interagency network involves more people entering a system, more contacts being maintained during different stages of treatment, and that these additional contact points result in greater difficulty in the maintenance of continuity. One other simple explanation is offered, (i.e., that continuity is a function of the size of the operation of a community mental health center).

Evidence of the extent to which community mental health centers are engaged in program evaluation activities related to continuity of care and assessment of efficiency is provided by the results of a survey of 325 federally funded centers in operation during July, 1971, which was conducted in the summer of 1972 by the Mental Health Services Development Branch of NIMH. A summary of the results of this survey prepared by Windle and Volkman (1973) indicates that of the 189 centers reporting, 61 percent indicated that they were engaged in evaluation activities concerned with monitoring the continuity of care within the center either for part of the center or the entire center; and 44 percent indicated that they were monitoring continuity of care between the center and other service programs for either part of the center or for the entire center. With regard to assessing efficiency, 38 percent indicated that they were engaged in evaluation activities for part or all of the center in terms of time and motion staff-time. Sixty-eight percent of those centers reporting in this survey indicated that they were engaged in evaluation activities for part or all of the center with regard to cost of care for the entire facility, while 54 percent indicated they were assessing efficiency with regard to cost of types of care. Although Windle and Volkman (1973) report that there is evidence that centers are using their program evaluation efforts to suggest changes for increased effectiveness and efficiency, these efforts are hampered by limitations in the resources available for such evaluation activities.

The NIMH Annual Inventory of Comprehensive Community Mental Health Centers does not provide data which indicate the extent to which centers are achieving continuity of care internally. However, considering the center as part of a total network of mental health service delivery components, the CMHC Inventory does provide some information as to the extent to which centers responding to this questionnaire helped to provide continuity of care in the total community mental health care delivery system. Specifically, one of the indicators derived from the CMHC Inventory is the percent of discontinuations from CMHC not referred and in need of further mental health care. As Table I shows, the percent of discontinuations in need of further mental health care and not referred to other caretakers increased over the four-year period from 1968 to 1971. Rather than showing progress, this data indicating an increase from 14 percent of discontinuations in 1968 to 22 percent in 1971 would indicate a regressive trend.

This regressive trend might be the result of a varying number of responding centers each year over the four-year period, and particularly the effects of the addition of new centers as respondents to the CMHC Inventory. However, data are also available from NIMH on a cohort of centers which control for the effect of new centers. This cohort of "The First Hundred Centers" provide longitudinal data which confirms this regressive trend. In "The First Hundred Centers" discontinuations not referred but in need of further mental health service increase from

18 percent of the total discontinuations in 1969 to 23 percent in 1971. Thus, there is evidence that an increasing percentage of clients are discontinued in terms of the services they received from the centers, are judged as still needing mental health care, but are not referred. Data on this cohort of centers also indicates that the number and percent of total discontinuations referred to other mental health care-givers, and to social and community agencies, declined from 1969 to 1971. NIMH is making further inquiry into this surprising finding.

TABLE I

*Percent of Total Discontinuities from Federally Funded Community Mental Health Centers not Referred Elsewhere and in Need of Further Mental Health Services**

Year	Percent Total CMHC Discontinuities
1968	14.0%
1969	19.3%
1970	23.5%
1971	22.4%

EFFICIENCY

Definition

In defining efficiency we deal with the ratio of output divided by input and ask the resultant question, "Can the same end result be achieved at a lower cost?" Katz and Kahn (1966) point out that there is a difference between potential and actual efficiency, and observe that there are "two quite separate aspects of the efficiency of any functioning system: the potential or abstract efficiency of a system design, and the extent to which that efficiency is realized in the concrete instance." These two aspects of efficiency, the potential for a particular organiza-

*Based on data included in Table 36 of NIMH Biometry Branch, OPPE, Survey and Reports Section report, "Descriptive Data on Federally Funded Community Mental Health Centers, 1971-72." Second Edition, May, 1973.

tional pattern operating under ideal conditions, and that which realistically can be expected from human beings under actual work conditions, while interdependent, must be recognized as involving somewhat different assumptions.

McLaughlin (1970) has observed that in one sense efficiency is a "dimensionless number", since it is the ratio of output to input and does not determine the volume of either input or output until one or the other is specified. Nevertheless, discussion of efficiency often produces a reaction of assuming that orienting a human service program to efficiency produces a net reduction in inputs and outputs. Whether this concern results in a lack of cooperation in service personnel's willingness to cooperate in providing cost data and data on service outcomes, it is sadly true that there is inadequate data on costs and outcomes of services to compare CMHC with respect to relative efficiency. It should also be mentioned that the benefits associated with service outcomes are particularly difficult to assess in the mental health field because of the low level of technological and conceptual development in this field.

Measurement

The problems of developing benefit-cost ratios to assess efficiency have been very difficult in health generally, although the economist and the systems analyst still consider benefit-cost ratios to be a basic professional tool. As McLaughlin (1970) has observed, because of the conceptual difficulties involved "in developing the universally accepted and politically acceptable benefit structure" there has been a tendency to shy away from formally accepting benefit-cost concepts and instead a "cost-effectiveness" approach has been employed. The cost-effectiveness approach seeks the most effective (usually lowest cost) method of achieving a given set of benefits.

We have indicated that full cost-benefit analysis is complex and difficult to achieve. Halpern and Binner (1972) have described a method which is simpler than a complete benefit-cost analysis which they call "output value analysis". Basically, the evaluative framework they propose involves estimating the economic value of a program's output by comparing its benefit value to the cost of achieving the output. It doesn't deal with all costs and benefits involved in the program but rather it focuses on two basic direct benefits of any mental health program and relates these to the immediate program cost.

A mental health program can produce a variety of products including custody for the dangerous, preventive intervention for a high risk group, consultation to community care-givers and return of patients to functioning in the community — this last product being perhaps one of the most important. Halpern and Binner suggest that two values may be attached to returning an individual to function in his community:

1) the value of the individual as an economically productive member of society;
2) the value that can be attached to his degree of improvement while receiving the services of the program.

In estimating the economic productivity of the patient or client, estimates of monthly earnings compared before service and after service can be used. Another major index of value is the degree of improvement obtained. Like the measure of economic productivity, the estimate or response value obtained from the service program is averaged across the range of clients receiving particular sets of services to deal with extreme values.

In estimating the costs incurred in providing service, Halpern and Binner (1972) suggest a procedure that allocates the cost of all of the indirect and supporting services to the direct services of the program, with the time that a client spends in each of the various statuses, such as inpatient, day hospital, family care, outpatient, etc. being multiplied by the cost of these statuses. In evaluating the program's functioning, they suggest using an output value divided by the estimated resource investment. In assessing effectiveness, it is suggested that the estimated output value would be equal to the estimated economic productivity plus the estimated response value (estimated response value is equal to the estimated response percentage multiplied by the average economic productivity).

Halpern and Binner also suggest indices for maximum response value, output value, and possible effectiveness. They have also developed a workload index which summarizes the direct services workload experienced by human service programs in terms of the input, process, and output components of the system. Halpern and Binner suggest that a single total workload index can be derived by combining number of client evaluations and admissions as an index of input work. They propose a process work-measure which is based on the number of clients enrolled in programs and the average number attending, and an output work-index which indicates the number of patients or clients discharged from the system. Preliminary results reported by Halpern and Binner indicate that output value analysis of this sort is useful in evaluating mental health programs.

The integration of services is itself a valued output of a comprehensive mental health and human service program and can also be examined as part of an output value-index related to the estimated resource investment required for achieving various levels of integrated service provision. One might improve on the model of output value analysis suggested by Halpern and Binner, by specifically adding appropriate indices of value or "benefits" received by clients as related to those values which would be expected to be increased by a well-integrated program. This type of data would be useful in comparing various administrative patterns for organizing services as they relate to outputs achieved at a client or patient level.

Research Evidence on CMHC Efficiency

There appear to be no major studies of the relative efficiency of community mental health centers at the present time. The work being done to develop output value analysis of mental health programs on an NIMH grant by Halpern and Binner (1972) as was mentioned earlier eventually may facilitate research on the efficiency of centers. However, at this time the data which is available seems to be limited to that gathered by the Biometry Branch at NIMH in its Annual Inventory of Community Mental Health Centers.

From 1970 to 1971 the average expenditure per person under care in community mental health centers decreased from $437 to $424, while the average number of persons under care per full-time equivalent staff person increased from 24 to 28. If one can assume that quality of care did not deteriorate, this data might be interpreted as indicating increased efficiency. Unfortunately, the requisite data on quality of care is not available.

Another indicator of the efficiency of operation of centers that may be derived from the Biometry Branch Inventory is the proportion of time spent by staff in indirect service, such as consultation and education, which reaches more persons per unit of effort and expands the staff person's impact. However, the percent of staff time devoted to indirect service in centers appeared to decrease from 11 percent in 1971 to nine percent in 1972, although such a small difference could result from errors in reporting or other unreliabilities in the Annual Inventory.

PATTERNS OF ORGANIZATION

Lacking sufficiently adequate data on continuity of care or efficiency, it is impossible to directly relate relative degree of accomplishment of these goals to the patterns of community mental health center organization. However, some problems of integration and continuity of care can be inferred from the "multi-organization" character of many community mental health centers. In an analysis of the forms of organization characteristic of community mental health centers, Levinson pointed out in a paper published in 1969 that at that time, three-quarters of the centers were formed by the affiliation of two or more agencies. In 1972, Feldman indicated that 85 percent of all the federally funded community mental health centers were comprised of "several different organizations working together under written agreements to provide a coordinated program". Ozarin, Feldman and Spaner (1971) describe one community mental health center as composed of 18 different organizations. Clearly, in such a highly differentiated community mental health center, the development of effective mechanisms for integration and for continuity of care becomes increasingly complicated and difficult.

Administrative Sub-Systems

Although a key factor in realizing both the goals of continuity of care and efficiency of operation in community mental health centers is the adequacy of administrative mechanisms employed by this type of organizational system to ensure integration of various components and to monitor input/output relationships, there is very little empirically-based knowledge about administration or management of any human service systems (Feldman, 1971). In introducing the new journal, *Administration in Mental Health*, Betram S. Brown, Director of NIMH, has pointed out the importance of adding useful literature in this area: "As mental health services have become more complex, there has been an increasingly apparent need for administrative processes that formulate cohesive programs and enhance public accountability." Clearly, research and development work is necessary not only in improving methods of assessing continuity of care, efficiency, and other process goals of community mental health centers, but also comparative research needs to be undertaken to identify effective patterns for organization and administration of community mental health centers.

FOOTNOTE

[1] Prepared for presentation at Symposium: "How Well is the Community Mental Health Centers Program Achieving its Goals?" American Psychological Association Annual Convention, Montreal, Canada, August 28, 1973. Preparation of the paper was supported in part by NIMH Grant MH 18382.

REFERENCES

Bass, Rosalyn D., & Windle, C. "Continuity of care: An approach to measurement." *American Journal of Psychiatry*, 1972, 129, 197-201.

Bass, Rosalyn D., & Windle, C. "A preliminary attempt to measure continuity of care in a community mental health center." *Community Mental Health Journal*, 1973, 9, 53-62.

Brown, B. S. "Introduction." *Administration in Mental Health*, 1972, 1, 1.

Feldman, S. "Problems and prospects: Administration in mental health." *Administration in Mental Health*, 1972, 1(1), 4-11.

Glasscote, R. M., Sussex, J. N., Cumming, Elaine, & Smith, L. H. *The Community Mental Health Center: An Interim Appraisal.* Washington, D. C.: Joint Information Service, American Psychiatric Association, 1969.

Gray, R. J. "Continuity of care in the community mental health center: Measurement and evaluation of selected administrative dimensions." Unpublished Master's Thesis, University of North Carolina, Chapel Hill, 1969.

Halpern, J., & Binner, P. R. "A model for an output value analysis of mental health programs." *Administration in Mental Health*, 1972, 1, 40-51.

Jepson, W. W. "Metropolitan mental health center development." In Grunebaum, H. (Ed.), *The Practice of Community Mental Health.* Boston: Little, Brown, 1970. Pp. 439-467.

Katz, D., & Kahn, R. L. *The Social Psychology of Organizations.* New York: Wiley, 1965.

Levinson, A. I. "Organizational patterns of community mental health centers." In L. Bellak and H. Barten (Eds.), *Progress in Community Mental Health.* Vol. I. New York: Grune and Stratton, 1969.

McLaughlin, C. P. "Systems analysis for health." In Sheldon, A., Baker, F., and McLaughlin, C. P. (Eds.), *Systems and Medical Care.* Cambridge: MIT Press, 1970. Pp. 230-267.

National Academy of Public Administration. "The multi-agency community mental health center: Administrative and organizational relationships." Final report on NIMH contract HSM-42-70-55, Accession No. PB-210-094, U. S. Commerce Department, National Technical Information Service, Springfield, Va., August, 1971.

National Institute of Mental Health, Biometry Branch, OPPE, Survey and Reports Section. "Descriptive data on federally funded community mental health centers." Rockville, Md., Second Edition, 1973.

Orden, S. R., & Stocking, B. B. "Relationships between community mental health centers and other care-giving organizations." Final report on NIMH contract HSM-42-71-7 by National Opinion Research Center, Accession No. PB-210-206, U. S. Commerce Department, National Technical Information Service, Springfield, Va., December, 1971.

Ozarin, L. D., Feldman, S., & Spaner, F. E. "Experience with community mental health centers." *American Journal of Psychiatry*, 1971, 127, 912-916.

Pugh, T. F., & MacMahon, B. "Measurement of discontinuity of psychiatric inpatient care." *Public Health Reports*, 1967, 82, 533-538.

Schwartz, M. S., & Schwartz, Charlotte G. *Social Approaches to Mental Health Care.* New York: Columbia University Press, 1964.

Socio-Technical Systems Associates. "Study of the relationships between community mental health centers and state mental hospitals." Draft final report on contract HSM-42-70-107, August 1972.

Wilder, J. F., Levin, G., & Zwerling, I. "Planning and developing the locus of care." In Grunebaum, H. (Ed.), *The Practice of Community Mental Health.* Boston: Little, Brown, 1970. Pp. 383-409.

Windle, C., & Volkman, Ellen. "Evaluation in the centers program." *Evaluation,* 1973, 1, 69-70.

As Baker and Broskowski note, community mental health centers have been regarded as "boundary busting" organizations. That is, centers have taken on the comprehensive mental health program responsibilities for particular geographic catchment areas, and then developed functions necessary to effect programs. In the process, community mental health staff have found themselves negotiating in, through, and out of a maze of human service systems. Whatever else, community mental health center staff have become experts on the organization and maintenance of agency boundaries. Not surprisingly, suggestions for new organizational forms for the human services have come, perhaps primarily, from persons involved with the community mental health movement.

The present paper attempts to delineate some alternative strategies for the development of organizational strategies in the human services that will both more realistically deal with the multi-problem family, without the problems created by elaborate referral processes, and result in an improvement in patterns of resources allocation and program effectiveness.

Frank Baker

Anthony Broskowski

CHAPTER 6

The Search for Integrality: New Organizational Forms For Human Services

HUMAN SERVICE BOUNDARY CHANGES

The boundaries of mental health and other human services have been changing over time as influenced by historical events, the changing power of vested interests, value changes in the larger society, the political situations of the times, and changes in professional ideologies and values. Periodically, all of these factors in interaction have given rise to a readiness to review boundaries of the service network and to reconceptualize a systemic pattern of service. In the 1960's the development of the community mental health perspective raised major questions about traditional care-giving systems (Kahn, 1969b; Schulberg and Baker, 1969).

The authors of one sociological analysis of the community mental health approach were moved to describe community mental health as a "boundary-busting system" (Dintz and Beran, 1971). It now appears that developments in the 1970's will further challenge the existing boundaries separating health, mental health and social services. Several recent publications have described the challenge of this decade as the development of comprehensive human service systems which will integrate traditionally separated services into comprehensive programs for human assistance (Demone, 1973; Schulberg, 1972; Schulberg, Baker and Roen, 1973).

Community mental health and human service approaches require the re-definition of boundaries defining the domains of both traditional organizations and professional disciplines. These boundary changes have primarily been of two types: boundary spanning and boundary expansion. In boundary spanning, emphasis is placed on the operational coordination of separate professional care-givers and care-giving agencies. In boundary expansion, the primary emphasis is on the incorporation of a wider variety of functions within a single organizational structure or, at the professional level, the development of generalist activities which cut across traditional disciplinary lines.

The movement toward developing coordinated and comprehensive patterns of service which cut across traditional organizational and administrative boundaries is apparent in the actions of many state governments to combine several separate health and social service programs in one new department. O'Donnell (1969) noted that by 1969 approximately twenty states had already combined several services and many other states are seriously considering similar reorganizations. In Massachusetts, for example, the new executive office of human services combines the traditionally separate public health, mental health, rehabilitation and social welfare programs of the state under one administrative control.

At the federal level, service integration legislation has been under development at HEW which would enable and encourage states and localities to unify the various programs and resources available to provide human services in order to facilitate the improved delivery and utilization of these services and thereby to increase their effectiveness in improving the lives of individuals and their families (*Washington Report on Medicine and Health*, January 10, 1972). The legislation would encourage, assist, and support state and local agencies in reorganizing, reassigning functions, and entering into new cooperative arrangements at different levels in the system of delivery of services. The federal goal is to meet human needs and alleviate or remove conditions of dependency as well as to improve the effective delivery of services. The integration of human services was initially invited in a bill, the Community Services Act, introduced in the Senate in 1967, at the request of Elliot Richardson, who was then Attorney General of Massachusetts. As HEW Secretary, Richardson was reported to have long favored the development of such legislation to provide integrated human services (*Washington Report on Medicine and Health*, January 10, 1972). Although at the time this is being written the passage of the Allied Services Act is still under consideration by Congress, HEW has also been reported to have gone ahead with over three million dollars worth of demonstration projects to test the feasibility of such programs (*Washington Report on Medicine and Health*, October 2, 1972).

Rationale for Service Boundary Changes

Why has there been such a widespread press for the re-definition of boundaries both in the community mental health movement and in other programs stressing unification of services? One aspect of the attraction is the desire to more efficiently and effectively allocate scarce resources. The spiraling costs associated with providing human services have certainly played a significant role in fostering a movement toward developing comprehensive integrative services which might synergistically combine previously separate health and social services programs.

Another major justification for integrated human services is recognition of the multifaceted nature of the problems faced by a family in their attempt to obtain the service resources that they need. Demone (1973), advocating the development of human services systems, reminds the defender of the status quo that he need only follow a few clients through the present network of services in order to realize that it is poorly designed to meet the needs of those it purports to serve. Demone writes, "a complex inter- and intra-organizational policy arrangement surfaces with boundaries which are often artificial and hamper the delivery of needed assistance. Such terms as fragmentation, overlapping, duplication, gaps in services, and lack of coordination take on concrete meaning for individuals who need services." Families with multiple problems often are able to receive help for only one of them, because the initial agency to which they turned for help either does not recognize the other problems or does not refer the family to other appropriate caregiving resources. Rein (1970) has also described and documented the problems of access for a family trying to secure needed services within a highly fragmented and specialized network in which each agency defines its own service boundaries:

> When a family needs a nursing home service for an aged member, remedial services for a child having difficulty with reading, or a summer camp, it is likely to encounter the problem of access . . . Access problems are exacerbated when a family has little income to purchase these services and must rely on the nonmarket system of public and voluntary social services.

New Organizational Structures

A major strategy for facilitating access to a comprehensive range of services has been to change the structure through which services are performed. Most attempts to improve the provision of community services tend to stress coordination of existing agencies and/or the extension of present programs. A questionnaire survey of Boston agencies by the authors confirms the impression that many agencies tend to see the primary needs of the community as being fulfilled by the expansion of their own agency. Recently, stimulated by the executive and legislative branches of government, there has been an increased press for the integration of comprehensive services within a single organizational entity. Community mental health centers, health maintenance organizations, neighborhood service centers, multi-service centers, neighborhood health centers, and youth opportunity centers are all examples of emerging organizations designed to accommodate this latter perspective. Demone (1973) reports that in January, 1971, HEW Secretary Richardson indicated that there were more than 2,000 such organizational units in the United States organized with the major goal of providing comprehensive integrated services.

In general, these newer organizational forms seek to provide improved assistance to clients to meet their mental health and other human service needs by incorporating the following features:

1. comprehensiveness of services;
2. decentralization of services in areas of high need;
3. concerting of resources from different programs;
4. co-location of service components;
5. operational integration of services in proper sequence thereby eliminating present duplication and wasted time for clients and employees (March, 1968).

Although these organizations share common goals there may also be important differences in their structure, management, and operating policies, depending upon their historical origins, the immediate environments in which they operate, and their relative emphasis on achieving internal coordination and integration of mental health and other human services. In general, these complex organizations evolve and operate without serious consideration of organizational definition or classification. There is presently a major need to provide a conceptual and empirically based definition and classification of these types of organizations. In developing useful organizational paradigms it seems necessary to begin with a substantial conceptual foundation closely related to a meaningful empirical base.

HUMAN SERVICES ORGANIZATIONS AS OPEN SYSTEMS

The authors and their colleagues have been developing a general *open systems* conceptual framework for the analysis of human service organizations. A brief summary of the approach will be presented in this paper and the reader is referred to a more detailed exposition in earlier papers (Baker, 1972; Baker, 1973a; Baker, 1973b; Baker, Broskowski and Brandwein, 1973; Broskowski and Baker, 1973; Schulberg and Baker, in press).

Conceptualized as an open system, a human service organization is defined as a bounded, interacting set of components organized in subsystems continuously engaged in an input-throughput-output process with an external environment. The primary components of the multiple transactions are people, information, money, and material resources. Based on general systems theory (Bertalanffy, 1968; Miller, 1965, 1972), this view of organizations as open systems emphasizes the importance of examining the internal interdependencies of sub-system components as they relate to their immediate sub-environments, with the optimal goals of growth, adaptation, and survival.

Human service systems differ from industrial production systems on which much of the organizational systems theory and research has been developed. Unlike business and industrial organizations, human service organizations usually have a much closer relationship to their people-consumer-clients. In fact, in residential organizations the consumers may be considered as organizational members. Furthermore, until recently, human service organizations have lacked the clear feedback channels that are frequently a part of production systems. Lacking profit and market mechanisms, there has been a need to develop mechanisms of accountability for the human service organization and much effort has been expended in developing such feedback loops in recent years.

New complex human service organizations are typically multi-unit, multi-goal systems. Although one can define the primary task of a human service organization as serving the public interest, there are many public interests and these newer systems must relate to a number of different, sometimes conflictive, constituencies. Through its relationships with various constituencies the human service organization becomes highly internally differentiated. As its various sub-systems simultaneously perform multiple functions and pursue different sub-goals, the organization experiences major problems of integration. Integration is one of the major defining characteristics of a "system" because without a degree of integration among the components, there is no unified totality. Components without integration do not constitute a system but merely a collection of individual elements isolated from each other.

Human service organizations, like all systems, develop heterogeneous internal parts, including a variety of specific operating service programs and managerial sub-systems. The organization's component sub-systems become differentiated according to adaptations to different parts of the external environment as well as the requirements of its respective service technology, and the beliefs of the professionals who work there. Operationally, sub-system integration is determined by the levels of collaboration, exchange, and mutual understanding which are achieved within the boundaries of the total system. It is the administrative or management sub-systems that have the major responsibilities for integrating the operating service sub-systems with each other and with maintenance sub-systems.

An open systems model helps one to maintain a perspective on the critical importance of the organization's environment. The human service organizational system can be usefully viewed as engaging in multiple transactions with a number of input and output constituencies within the following major sectors:

1. clients and consumer groups (both potential and current users);
2. suppliers of sanction, staff, funds, materials, technology, and information;
3. other individuals, groups, and agency service providers who may compete or cooperate with regard to clients and resources; and
4. regulatory organizations, including governmental agencies and professional associations.

Criteria of Service Integration

Before presenting an overview of organizational structures for achieving integration of services, we would like to review some of the operational measures that are potential indices of integration. In developing criteria for integration it is important to distinguish at least two related issues:

1. the extent to which internal service components actively relate to one another, both in terms of general program activities, and the activities centered around an individual consumer.

2. the extent to which the organization's multiple services are integrated with those in its larger environment.

In examining the integration of internal components it is important to distinguish the effort which is devoted to achieve integration from the degree of integration that is actually achieved. In a comparative study of multi-unit organizations in industry, Allen (1971) successfully employed the total amount of time devoted by working managers to achieving integration as an index of integrative effort. In terms of the degree of integration achieved, it is important to examine the collaboration and achievement of mutual understandings among units. Collaboration might be assessed by examining participants in organizational decision-making, particularly as the decision-making relates to the distribution of scarce resources and the disposition of clients. One measure reflecting noncollaboration could be the degree of self-containment of various units in terms of activity patterns and the extent to which the unit exclusively depends on its own staff to accomplish various functions. One could also assess the types of inter-unit communication as well as the use of specific information exchange devices.

In addition to the exchange of staff and information, the exchange of clients and material resources is also of major concern. The number of jointly engaged projects including planning, training, and experimental treatment programs could also be relevant. The degree of cross-unit identification, the use of unifying identifying labels for multiple units, and a high degree of agreement about the respective roles of units would be indicative of a high degree of mutual understanding.

At the level of individual care patterns, it would be important to examine typical patterns of care as these relate to the range of various types of service units used by clients and the actual number of different sources of care that a client must employ. It would be expected that being served by a comprehensive integrated human service center would require a client to use fewer different organizational sources in seeking the total care that is needed.

Schumaker (1972) has developed measures of two major dimensions of individual care patterns — compactness and cohesiveness. Compactness of an individual care pattern is determined by the range of sources or types of sources of care used by a client or consumer. The smaller the range, the more compact is the pattern. The range of this dimension exists on a continuum from a single course of care to multiple sources of care. Schumaker defines "single type of multiple sources" as a situation in which a consumer uses a number of different sources of care which are of the same type, including groupings such as several emergency rooms or several outpatient departments. The category of "multiple type of sources" refers to situations in which a consumer makes use of a number of different types of health and social care-givers. Such a pattern would involve the use of services at an outpatient department of a community mental health center, the emergency room of a general hospital, a private physician, welfare agency, and so on.

Schumaker also proposed an index of the "cohesiveness of a care pattern" which he defined as "the degree of integration of sources of care into an organized pattern (Schumaker, 1972). The cohesiveness of a human service care pattern is related to the number of sources of care in the pattern, to the number of sources of a particular type in the pattern, and to the number of types of sources in the pattern. This index would consider such dimensions as duplication, supplementation, completeness, and dispersion. This index also deals with the question of whether the patient or client is using more services than appear to be necessary. Other questions of concern are whether there are sufficient sources to meet the patient's needs and those of his family and whether there are obvious gaps and discontinuities.

Closely associated with the degree of effective integration at the level of the client as well as between organizations, is the concept of service continuity. The degree of concert achieved among organizational components and the extent to which there are adequate service linkages will affect the degree to which services become discontinuous and a client fails to receive necessary simultaneous or sequential services.

In reference to community mental health centers, Bass and Windle (1972) have defined continuity of care as operationally existing to the extent that "there are no obstacles to a client's either remaining in or moving from any of the center's direct treatment services in conformity with his therapeutic needs," and to the extent that "administrative mechanisms relate past and present care by providing stable client-caretaker relationships, necessary written and verbal communications among staff members, and contact with clients who miss appointments, go on unauthorized leave, or otherwise appear to be dropping out of treatment prematurely."

Their definition highlights four criteria of continuity:

1. client movement or the absence of it in appropriate response to treatment needs;

2. stability of the client-caretaker relationship;
3. easy flow of both verbal and written communication among staff members;
4. efforts made to retrieve clients who appear to be dropping out of treatment prematurely.

Bass and Windle have developed a five-part inventory designed to be used during a limited observation period to measure the continuity of care between the direct essential services of a community mental health center. Because of the requirements for long-term on-going data collection the Bass and Windle inventory tends to have somewhat limited applicability unless the organization being examined is willing to institutionalize such measurement. Nevertheless, Bass and Windle (1972) have performed a useful service by attempting to operationally define a key concept useful in judging the performance of a comprehensive service center.

Operational measures of service integration offer a potential procedure for testing the adequacy of new organizational forms for human services. Unfortunately, organizational arrangements developed with the goal of achieving integration have rarely been tested by the application of such empirical methods with the result that service integration tends to remain a somewhat abstract concept.

ORGANIZATION AND ADMINISTRATION OF INTEGRATED SERVICES

With these indices of integration in mind, we would now like to shift our focus to the various organization forms and administrative mechanisms that are presently emerging for the delivery of human services. Although effective management is a key factor for achieving integrated human services, there is presently very little empirically based knowledge concerning the administration of human service systems (Feldman, 1972). In introducing the new journal, *Administration in Mental Health*, Bertram S. Brown, Director of NIMH has pointed out the importance of adding useful literature in this area: "as mental health services have become more complex, there has been an increasingly apparent need for administrative processes that formulate cohesive programs and enhance public accountability" (Brown, 1972).

Our review of organizational forms and administrative mechanisms will begin with those presently existing organizations that are tending toward integrated human services and conclude with those structures which are only now beginning to emerge. Each form has its advantages and disadvantages and each will be adopted at the expense of various trade-offs.

Community Mental Health Centers

There are currently 515 community mental health centers (CMHC) funded by the National Institute of Mental Health, of which almost 350 are already in operation (Brown, 1972). In analyzing the organizational forms of CMHC, Levenson (1969) pointed out that in 1969 three-quarters of the centers at that time were formed by the affiliation of two or more separate agencies. Feldman (1972) indicated that by 1972, 85 percent of all the federally funded CMHC were comprised of "several different organizations working together under written agreements to provide a coordinated program." Ozarin, Feldman, and Spaner (1971) describe one CMHC as composed of 18 different organizations.

This organizational model can be considered as an example of a "multi-organization" [i.e., a union of parts of a number of organizations which come together for the performance of a common task (Stringer, 1967)]. In a highly differentiated multi-organization such as this type of community mental health center, the development of effective mechanisms for integration becomes increasingly important.

Community mental health centers include elements closely related to traditional medical services as well as those more closely allied to social services. While some community mental health centers appear to be sticking to a traditional organizational model of the delivery of medical care, others are beginning to initiate a move away from the medical model of care to a human services model which in some ways is closer to social service agencies. Some centers we have visited are closely tied, through contracts and other agreements, to a wide variety of planning, service, and research related agencies. The degree to which centers will deviate from the more segregated medical model of mental health will depend on their organizational origins and auspices, as well as the density of human service programs in their own immediate environments.

Perlmutter and Silverman (1972) have asserted that community mental health centers whose parent bodies were general hospitals, medical centers, and state hospitals, tended to pattern their organizational structures after the medical institutions from which they were developed.

Smith (1968) had earlier warned that community mental health centers that were hospital-centered would tend to follow a more traditional medical model under medical control and that this would tend to minimize the development of innovations and the offering of truly comprehensive services.

Our own experiences in participating in evaluative site visits to NIMH-funded community mental health centers also indicates that those centers situated in isolated areas lacking strongly established social agencies and having few medical institutions tend to develop more comprehensive and integrated social service programs (Socio-Technical Systems Associates, 1972). Centers in "service dense" urban areas, where other agencies compete for turf and clientele, tend to restrict themselves to the basic mental health services. Exceptions, of course, are readily available.

Models of CMHC Administration

The administration of community mental health centers also shows wide variations. While we know of no comprehensive research or survey on this matter, we have identified some interesting differences and have noted in the literature some distinctions we felt would aptly apply to center administration.

Pusic (1969) has identified two models of administration — territorial and functional. These two models represent end points on a continuum with most organizations having some elements of both. Pusic begins by defining administration as "the continuing coordination of socially accepted activities aimed at the realization of possibly diverging interests."

Territorially administrated organizations, according to Pusic (1969), have five distinguishing characteristics. They are organized around a geographic and/or political base. They are multi-functional, encompassing several technically unrelated tasks. They accomplish complex objectives through successful implementation of smaller divisions of labor, until specific tasks are assigned to individuals. A pyramidal hierarchy is used to coordinate the operation with "a chain of command, ladder of advancement, and a line of reporting and subordination". Conflicts are resolved by power. Priorities exist to service and interests of owners, political or professional power groups, or providers of funds.

In contrast, Pusic (1969) describes functionally administrated organizations as having the following characteristics: They tend to be organized around a single social function or technically interrelated functions; the knowledge and skills of separate specialists are integrated into progressively wider and more complex combinations of technically determined functions; there is a less hierarchical distribution of power throughout the system, with linkages achieved through technical interdependencies, professional norms, and shared activities. There is, consequently, less potential for conflict. The interests of members are more common, overlapping, and stable than in the territorial administration model because such functionally-centered organizations are limited in their ability to choose among ends as well as among means.

Perlmutter and Silverman (1972) suggest that the original planning for community mental health centers called for the implementation of a model which was closer to the pattern of functional administration than to the more traditional territorial model. On the basis of an examination of four community mental health centers in Philadelphia, however, Perlmutter and Silverman questioned whether the functional alternative has been generally implemented in community mental health centers.

Our own experience has generally indicated that management through differentiation by professional departments (e.g., psychiatry, nursing, etc.) is stronger in centers operating in affiliation with universities for purposes of training and research. It is probably true that the training and research functions reinforce the specialist or professional reference group identification, and make coordination across direct service units more difficult. Such centers are also likely to thereby experience difficulties in the integration of services from the perspective of the multi-problem client. Organizational differentiation by priority tasks, client sub-groups, or special services, is more likely to occur in centers that are not dominated by a medical model or professional hierarchy, and are likely to be found in areas of professional scarcity or where large new population expansion has taken place.

Multi-Service Centers

Another emergent organizational form with the goal of providing comprehensive and locally responsive services is the multi-service neighborhood center. Initial efforts centered around the creation of new neighborhood centers which could coordinate the social services and human resource programs funded under the Juvenile Delinquency-Youth Offenses Control Act of 1961. Recognizing that delinquency stems from a complex set of interrelated causes, federal demonstration grants were made to a number of cities to develop comprehensive programs, including neighborhood centers. The broader efforts of the "war on poverty" overtook these programs and gave neighborhood centers further impetus for comprehensive expansion (March, 1968). The Economic Opportunity Act of 1964 provided authority to OEO to fund hundreds of neighborhood centers as part of a broad spectrum of programs attempting to deal with problems of poverty.

Neighborhood service centers were developed to achieve, among other goals, the integration of previously fragmented services. Perlman and Jones (1967) described the situation with which neighborhood service centers were attempting to deal:

> Problems were interrelated, but relief was to be found only if the client could piece together services organized according to the specific functions of agencies.

The multi-service center usually includes a community organization component and is frequently committed to social-action. The few available studies of neighborhood service centers generally agree that their presence has made more services available to the local community (Kirschner Associates, 1966; Perlman and Jones, 1967; O'Donnell and Sullivan, 1969). However, as Kahn (1969) summarized it, "data about effectiveness, efficiency, and innovation were mixed;" and he observed that the limitations of those centers studies were traceable to a "conflict as to whether the centers housing local programs were to be judged in terms of service output, development of local leadership, or employment of poor people."

March (1968), in analyzing the neighborhood center concept, has described several neighborhood center models for organizing service delivery systems. In the comprehensive model, the "One-Stop, Multi-Purpose Neighborhood Service Center" is a comprehensive array of mental health, physical health, family life activities, education, manpower, housing, recreational and legal services. These services are co-located at one neighborhood center. March describes the goal of this type of neighborhood center as gathering together the service agencies of the neighborhood and concerting their activities for human resource development as a "single system in providing preventive therapeutic and rehabilitative services and assistance to restore the handicapped to social and economic self-sufficiency." The advantages of the "one-stop" model are difficult to achieve, but centers in a number of large cities are attempting to develop such a comprehensive integrated structure for services. Others have also noted that such multi-service centers may become internally fragmented over conflicts of direct services versus community organization.

Neighborhood service centers have experienced major decreases in the anti-poverty monies which originally encouraged their growth. A number are surviving, however, and are still attempting to implement improved models of service. Similarly, community mental health centers will eventually have to get along without major federal support. Both the mental health center and the multi-service neighborhood center will have to function with multiple and integrated funding sources. Nader's Task Force on NIMH (Chu and Trotter, 1972) has suggested, for example, that a national social service insurance, independent from, but parallel to, national health insurance, be set up to finance the wide range of human service programs necessary in the modern community.

Consortiums and Network Models

Another model, currently less prevalent, but becoming more common, is the development of a network of existing agencies, contractually tied together, to provide a system of integrated services. This model is based on the rationale that there are not enough resources to create new organizations and agencies and that the predominant goal is to more effectively utilize what is already in existence. This viewpoint appears to be evolving in federal policy and was specifically mentioned by Richardson (1973) in his final report on HEW. This network model, however, doesn't simply rely on the development of "consensus coordination or gentlemen's aggreements". Rather, it works on the principle of explicit contracts calling for reciprocal exchanges of money, staff, space, resources, and clientele.

While difficult to achieve, if properly constructed, such consortiums are very effective at a minimum of additional costs. Modern techniques of cost-accounting, coupled with innovative enabling legislation, have allowed for a variety of such integrated service networks. One example familiar to the authors, is the extremely complex set of contracts between over a dozen agencies in Tucson, Arizona, developed to provide integrated human services to the elderly. The funds come from as many as eight sources and the exchange of staff, space, and clientele is extremely complex but well operated. Furthermore, it is not possible to identify a single organization that tends to control the flow of funds, staff or clients. Unlike the multi-organization community mental health center model or multi-service neighborhood center model, the consortium model is not focused or controlled by a single corporate entity. Like any complex adaptive system, it is very difficult to locate in a single controlling unit. Rather, a network of functions, with appropriate checks and balances, tends to respond to meet client-centered problems.

The administrative arrangements take a great deal of time to develop and can be extremely difficult to conceptualize and comprehend. Our interviews with clients and service staff of the Arizona agency network, however, did not seem to indicate that they cared much about the contractual complexity as long as they were getting all the necessary services without encountering such barriers as conflicting ideologies, eligibility requirements, payment schemes, and service discontinuities.

Information and Referral

A final organizational arrangement that can be used to integrate services is the well known but frequently under-utilized information and referral center. Nichalous Long (1973) has recently written about information and referral (I&R) services in considerable detail. This mechanism has the advantage of being relatively cheap and easily developed. Its disadvantages are that it does not directly address the issue of guaranteed continuity of services nor the development of missing services. Agencies may not accept referred clients or there may be gaps in service for particular types of problems. An I&R mechanism, is, nevertheless, relatively effective as an immediate way to begin integrating multiple services in a service-dense area. I&R services can also provide the valuable functions of monitoring on a continuous basis the changing needs of the people in the community and providing planners with data on relative priorities or geographical areas of need. The I&R model can also be used in conjunction with, or within the context of, one of the previously mentioned organizational forms. For example, Bloom (1972), has described the use of an elaborate and computer-based I&R system used within a consortium of agencies linked to a community mental health center.

Very little is presently known about the particular administrative processes that are vital to I&R effectiveness. Broskowski and Baker (1972) have provided a case illustration of some of the functions and difficulties of an I&R service within the broader organizational context of a health and welfare organization. I&R administration appears to require a strong generalist orientation with a real flair for interagency relationships and client advocacy. It is often the case that the person that must resort to an I&R service to locate help has already exhausted obvious sources of services or is so isolated from natural support systems that he does not know where else to turn. Frequently the I&R center is the court of last resort. Its management and staffing must reflect this fact.

THE NEED FOR SYSTEMATIC STUDY OF NEW SETTINGS

Sarason (1972) has observed that in the past decade or so, more new "settings" (a word which he views as interchangeable with programs, organizations or institutions) have been created than during the entire previous period of human history. We concur with Sarason's point that an increasingly common approach when attempting to solve a whole range of social problems is to create new organizational forms.

Many of these newly created organizations are mandated to deliver and coordinate the essential components of service that are necessary to meet complex human needs. We have described some of these integrating organizational structures, such as the multi-service center and the community mental health center, as well as such inter-organizational arrangements as consortiums, agency networks, and information and referral services. Unfortunately, these and many other new organizational arrangements and structures rarely have an adequate testing before further new "settings" are developed. The search for integrality in human services will continue and if the social arrangements which are developed to meet this goal are to be more than the embodiments of evolving political philosophies and professional ideologies, close empirical examination of the systemic character of these social experiments must be systematically carried out. We hope that such human service organizational research and development will be presently forthcoming.

REFERENCES

Allen, S. A. "A comparative analysis of corporate-divisional relationships in multi-unit organizations." Paper prepared for meeting of the Institute of Management Sciences, College on Organization, March 22-24, 1971, Washington, D. C.

Baker, F. "Planning and the environment of a community mental health center." *Psychiatric Quarterly*, 1972, 46 (1), 95-108.

Baker, F. "Organizations as open systems." In F. Baker (Ed.), *Organizational Systems: General Systems Approaches to Complex Organizations.* Homewood, Illinois: Richard D. Irwin, 1973a.

Baker, F. "The living human service organization: Applications of general systems theory and research." In H. W. Demone and D. Harshbarger (Eds.), *A Handbook of Human Service Organizations in Context.* New York: Behavioral Publications, 1973b.

Baker, F., Broskowski, A., & Brandwein, R. "System dilemmas of a health and welfare council." *Social Science Review*, 1973, 47, 63-80.

Bass, R. D., & Windle, C. "Continuity of care: An approach to measurement." *American Journal of Psychiatry*, 1972, 129, 196-201.

Bertalanffy, L. von. *General Systems Theory.* New York: George Braziller, 1968.

Bloom, B. L. "Human accountability in a community mental health center." *Community Mental Health Journal*, 1972, 8, 251-260.

Brown, B. S. "Introduction." *Administration in Mental Health*, 1972, 1, 1.

Broskowski, A., & Baker, F. "The functions of an information and referral subsystem in a community planning organization." Laboratory of Community Psychiatry, Harvard Medical School, Unpublished paper, December, 1972.

Chu, F., & Trotter, S. "The mental health complex; Part I, Community mental health centers." *Task Force Report on the National Institute of Mental Health.* Washington, D. C.: Center for Study of Responsive Law, 1972.

Demone, H. "Human services at state and local levels and the integration of mental health." In G. Caplan (Ed.), *American Handbook of Psychiatry.* Vol. 2. New York: Basic Books, 1973.

Dintz, S., & Beran, N. "Community mental health as a boundaryless and boundary-busting system." *Journal of Health and Social Behavior*, 1971, 12, 99-107.

Feldman, S. "Problems and prospects: Administration in mental health." *Administration in Mental Health*, 1972, 1(1), 4-11.

Kahn, A. J. *Studies in Social Policy and Planning.* New York: Russell Sage, 1969.

Kirschner Associates. *A Description and Evaluation of Neighborhood Centers.* A report for the Office of Economic Opportunity, December, 1966.

Levenson, A. I. "Organizational patterns of community mental health centers." In L. Bellak & H. Harten (Eds.), *Progress in Community Mental Health.* Vol. 1. New York: Grune & Stratton, 1969.

Long, N. "Information and referral services: A short history and some recommendations." *Social Service Review*, 1973, 47, 49-62.

March, M. S. "The neighborhood center concept." *Public Welfare*, 1968, 26, 97-111.

Miller, J. G. "Living systems: Basic concepts." *Behavioral Science*, 1965, 10, 193-237.

Miller, J. G. "Living systems: The organization." *Behavioral Science*, 1972, 17, 1-182.

O'Donnell, E. *Organization for State Administered Human Resource Programs in Rhode Island.* Report to the General Assembly by the Special Legislative Commission to Study Social Services. June, 1969.

O'Donnell, E. J., & Sullivan, M. M. "Service delivery and social action through the neighborhood center: A review of research." *Welfare in Review*, 1969, 7, 1-11.

Ozarin, L. D., Feldman, S., & Spaner, F. E. "Experience with community mental health centers." *American Journal of Psychiatry*, 1971, 127, 912-916.

Perlman, R., & Jones, D. *Neighborhood Service Centers.* Department of Health, Education, and Welfare. Washington, D. C., 1967.

Perlmutter, F., & Silverman, H. A. "Community mental health centers: A structural anachronism." *Social Work*, 1972, 17, 78-84.

Pusic, E. "Territorial and functional administration in Yugoslavia." *Administrative Science Quarterly*, 1969, 14, 62-72.

Rein, M. *Social Policy: Issues of Choice and Change.* New York: Random House, 1970.

Richardson, E. "Responsibility and responsiveness (II), A report on the HEW potential for the seventies." Department of Health, Education, and Welfare, Washington, D. C., 1973.

Sarason, S. S. *The Creation of Settings and the Future Societies.* San Francisco: Jossey-Bass, 1972.

Schulberg, H. C. "Challenge of human service programs for psychologists." *American Psychologist*, 1972, 27, 566-573.

Schulberg, H., & Baker, F. "Challenge of human service programs for psychologists." *Psychiatric Opinion*, 1969, 6, 14-26.

Schulberg, H., & Baker, F., & Roen, S. (Eds.), *Developments in Human Services*, Vol. I. New York: Behavioral Publications, 1973.

Schulberg, H., & Baker, F. *The Mental Hospital and Human Services.* New York: Behavioral Publications, in press.

Schumaker, C. J. "Change in health sponsorship: II. Cohesiveness, compactness and family constellation of medical care patterns." *American Journal of Public Health*, 1972, 62, 931-935.

Smith, M. B. "The revolution in mental health care — A 'bold new approach'?" *Transaction*, 1968, 5, 19-23.

Socio-Technical Systems Associates. *A Study of the Ways in Which Community Mental Health Center Services are Integrated into Other Types of Care-Giving Functions.* A report for the National Institute of Mental Health, August, 1972.

Stringer, J. "Operational research for 'multi-organizations'." *Operational Research Quarterly*, 1967, 18, 105-120.

Washington Report on Medicine and Health, January, 10, 1972.

Washington Report on Medicine and Health, October 2, 1972.

Harold W. Demone, Jr.

If adaptive changes are to be effected in the organization of human service systems, it will be necessary for us to pay far greater attention to human service related policy decisions than we have done in the past; in addition, it will be necessary to generate effective strategies for influencing policy decisions. Unfortunately, the formal training, past experience, and guild sanctions of mental health and other human service professionals have only occasionally equipped them to operate in this complex political arena.

In discussing some of the roles people play in influencing policy, Demone focuses on four operational strategies: change agent, lobbyist, ombudsman, and citizen participation, and discusses the assets and liabilities of each strategy. These strategies are presented as roles which the human service professional might simultaneously play, although in different arenas and with different constituencies, as well as roles he might orchestrate and coordinate in the policy influencing behavior of his associates.

Whatever one's feelings about engaging in the political process of influencing policy decisions, the human service professional must view this problem seriously. For, it is in these decisions that the basic conditions affecting an organization's operation and survival are established. We ignore them at our own peril.

CHAPTER 7

Influencing Policy: Some Roles People Play

How does an individual, group or organization effectively make an impact upon the policy-making process? Conceptually, theories of social systems, general systems, social action, politics and futurism all offer a frame by which understanding can be enlarged. An understanding of the roles and types of major actors and participants in the policy engagement is also helpful. The individual may act as a citizen, a service consumer, or a provider; he can represent himself, an informal group, or a formal organization. Some of the roles which can be played (among many others) are change agent, lobbyist, ombudsman, consumer and citizen, which will be discussed further in this paper.

Before beginning on elaboration of these policy-influencing roles, some additional considerations and definitions need explication. The change maker has available a variety of change processes: community organization, community development, bargaining, consensus seeking, consultation, forums, conferences, coalitions, seminars, advocacy, contest, conflict and militancy. The available technology includes self-study, information systems, operational research, program-planning-budgeting systems and long-range planning. Thus, there emerges an overlapping matrix of reasonably well-defined roles played against a series of change processes using a variety of technologies to achieve certain policy ends.

What is public or private social policy? Since policy decisions may range from a New Year's resolution to a decision to fly to the moon, room for discussion and analysis is possible. And since we reached the moon and then decided to reduce our space program; and often break our resolutions, it should also be clear that policies are only sometimes achieved and even then are often fluid. Thus policy should not be regarded as immutable and unchanging, but rather as dynamic and constantly in tension.

> Social policy may be regarded as the principles and procedures guiding any measure or course of action dealing with individual and aggregate relationships in society. It is conceived as intervention in, and regulation of, an otherwise random social system. Established social policy represents a settled course of action . . . (Schorr and Baumheier, 1971).

To the National Association of Social Workers (1959), public social policy:

> . . . consists of those laws, policies, and practices of government that affect the social relationship of individuals and their relationship to the society of which they are a part. Such policy operates directly through public social programs and indirectly through its influence on voluntary activities and relationships.

The American Heritage Dictionary (1969) defines policy as:

> any plan or course of action adopted by a government, political party, business organization, or the like, designed to influence and determine decisions, or procedures considered to be expedient, prudent, or advantageous . . . Prudence, shrewdness, or sagacity in practical forms.

Medical, psychiatric and sociological dictionaries ignore policy as apparently not sufficiently specific to be included in their specialities. Political scientists revel in it, and it has become an important concern to social workers. Thus, individual disciplines view policy differently, and professions that do treat policy as a legitimate concern tend to see it as principally a function of government.

Yet, as we noted earlier, people make policy decisions acting in their ascribed roles in organized collectivities, and as individuals in their relations with others. As such they may be public or private. Ostensibly representing all the people, it is true that public decisions in a public forum may have a significant lasting effect on many people. But, the cumulative results of many private policy decisions may be equally important. Not surprisingly, one of the principle functions of our complex public regulatory fabric is to limit private policy decisions. Since there is no way to measure and contrast the relative influence of the public and private sectors, the importance of such distinctions must be considered conjectural, but they need to be made. To be sure, policy is not the exclusive property of the public sector.

The private policy resulting from negotiations between the United Auto Workers and General Motors in the mid-1960's which substantially stimulated third-party coverage of mental illness was a highly significant step in returning mental illness to the health umbrella and legitimizing treatment of those so afflicted. Certainly, the decisions of Rockefeller, Carnegie, and later, Ford, to establish foundations directed to the public good has led to many changes in our society. The oil shortage crisis of 1973 was attributed to the policies of eight large American oil companies; and ITT in the same year was accused of heavy-handed influence on smaller South American countries. Marxian economists, in fact, would stress the weakness of public policy in a capitalistic state.

Debate about public policy underlines the inexorable linkage between public and private, profit and non-profit, and the individual and society. The 1971 Department of Health, Education, and Welfare White Paper on Health (May, 1971) enunciates in the foreward, by then-Secretary Elliot L. Richardson, a clear public policy:

> . . . given the choice between extending the activities of the Federal Government, and using the forces of the private sector to achieve an objective, the latter was preferred . . . Preference for action in the private section is based on the fundamentals of our political economy — capitalistic, pluralistic and competitive — as well as upon the desire to strengthen the capability of our private institutions in their efforts to provide health services, to finance such services, and to produce the resources that will be needed in the years ahead. (HEW, 1971)

The gauntlet is clearly defined; for example, the present Administration would retain the giant insurance industry (both Blues and commercials), while the opposition would have the government assume this role. In any case, both the debate itself and the subsequent actions will have a lasting effect not only on the health of individual Americans but on the complex health industry, government and private, profit and non-profit. To successfully achieve impact on these and other policy decisions is the goal of many individuals and organizations. Of the many alternative roles which can be played, four — change agent, lobbyist, ombudsman and citizen participation — will be discussed.

THE CHANGE AGENT

The increasing popularity of Saul Alinsky until his death in 1972, the dynamics of a Ralph Nader, the influence of Harvard child psychiatrist, Robert Coles, and the work of Cesar Chavez are all evidence of the public effectiveness of identifiable change-seeker role models.

Against an exponential rate of technological change our society and its institutions, law, religion, government, the family and the professions, are increasingly placed in stressful positions. For example, professionals inquire as to their proper roles: Should the physician hang up his shingle as a solo practitioner, join a group of his peers, become a full-time hospital staff member, or work in a neighborhood health center in a low-income community?

Edgar Schein (1970) suggests a three-fold typology for professionals. *Custodianship professionals* accept fully the norms of their group, the current knowledge level, and skills. They are satisfied with the status quo. They favor strong professional associations. The *content innovator* accepts the professional norms but wants to see improvement in knowledge and skill. He concerns himself with scholarship, knowledge and science. He is less likely to engage in practice, and instead involves himself in teaching and research, and views his academic colleagues as his peer reference group. He is considerably less involved in, and often indifferent to, professional associates unless they threaten his career. His goal is to maximize technical competence. The *role innovator* is rare and is differentiated from the content innovator because he not only challenges the technology but also the technocracy, norms and values of the profession.

> . . . he questions the traditional ideas as to who is a legitimate client, who can or should initiate the contact between client and practitioner, what constitutes an appropriate setting for conducting professional activities; and what are the legitimate boundaries of the professional's area of expertise.

Ralph Nader is described by *Time Magazine* (1969) as "The U.S.'s Toughest Customer" and "an almost legendary crusader". He sees the law as a powerful instrument for positive social change. All consumers are his clients even though few have requested his assistance. To Nader, legal services are a right, not a privilege. He has been principally responsible for the enactment of six major federal laws — the National Traffic and Motor Vehicle Act of 1966, the Wholesome Meat Act of 1967, the Natural Gas Pipeline Safety Act, the Radiation Control for Health and Safety Act, and the Wholesome Poultry Act, all enacted in 1968; and in 1969, the Federal Coal Mine Health and Safety Act. His other targets have been equally varied: the use of monosodium glutamate in baby food, fatty hot dogs, unclean fish, tractors, medical x-rays, color television radiation emission, a major Washington law firm, and many federal agencies.

Another change agent of note was Saul Alinsky who spent more than thirty years organizing the poor only to move, late in his life, to organizing the middle class. His premise was that all people had power if they developed political and organizational skills. *Time Magazine* (1970) in an essay devoted to Alinsky, suggests that he "has possibly antagonized more people — regardless of race, color, or creed — than any other living American."

To Alinsky, conflict was the road to progress; sharing of power, the essence of democracy. His targets were most often city officials, corporations, and slum landlords. His tactics included boycotts, pickets, rent strikes, and a variety of imaginative stunts. According to Alinsky, to develop credibility a group needs an enemy, a target of adequate stature. Eastman Kodak Company and the University of Chicago proved to be admirable opponents in the 1960's (*Time*, 1970). In his last book, published in 1971, his sophistication was most apparent. "You never take an action without first figuring out the reaction. Periodic mass euphoria around a charismatic leader is not an organization." And finally, "He who lives by the sword shall perish by the champagne glass." (Alinsky, 1971)

These highly charged change agents are not limited to the national scene. Each state and area can usually identify someone who plays these types of roles around various issues. For example, *Harper's Magazine* featured James R. Ellis, one of the State of Washington's most influential men. He has never held public office, nor has he aspired to it. He is not wealthy. He is not a member of the upper class. He is not charismatic. He is not a part of a major political machine. He is the opposite of the power elite. Yet, according to *Harper's* (1969), he is substantially responsible for much of what has happened in that state's public arena for the last decade.

Ellis, a lawyer, is well organized. He can array the facts and clearly set forth his ideas. He is listened to although such was not his fate when he began. It was only after seven years of failure in trying to attract support for an effective regional government for Greater Seattle that he turned to focused, incremental change. He found that people were willing to tackle a single dramatic project. He was successful and he never showed any personal political ambition. His credibility began to grow.

Unlike Nader and Alinsky, he used consensual methods, arraying broad support. He was successful in specific, identifiable civic improvement programs, unsuccessful in complex, devisive issues.

A final example of a contemporary change agent is Harvard Psychiatrist, Robert Coles. Through his thirteen books and more than 350 articles, he has worked to communicate the message that most poor whites and blacks, and working-class whites, are healthy in mind and courageous of spirit. According to Coles, they are possessed of untapped strengths that if utilized and understood, could be most important for America (*Time*, 1972). Consulted by congressmen, studied carefully by the academic community, featured in a cover story by *Time*, he has been quoted frequently by the nation's popular press. *Time* calls him "the most influential living psychiatrist in the U.S." To Kenneth Clarke, Coles "keeps morality, decency, and justice alive."

Four exciting, dynamic innovators, all different, using different change strategies and tactics, yet all successful, have been described. Thus, although the ends may be one, the means are many. The setting, structure, skill, personality, flexibility, ability, and interest of the change agent, working in a complex, rapidly changing environment, are all variables of note and importance. Although it is possible to limit oneself to a single change tactic, success over time is likely to be limited.

Change agents need not have been professionably committed to social change over time. The 1950's and 1960's saw many fluoridation battles across the United States, and dentists led the way. Since dentists are not known, as a group, for their role-innovating behavior, this was a most important example of an issue finding advocates. Irwin Sanders (1969) says, "Many dentists have been thought of, and many thought of themselves, as repairmen rather than medical men . . . They were said to be more interested in techniques of filling teeth than of viewing caries as a disease." Yet they assumed a public-political role having the customary successes and suffering the usual failures common to the political arena. Given the right issue and the right time, individuals (and professionals) ordinarily removed from the public will risk their image and participate in political processes.

And there are professional change agents as well. Part of well-organized bureaucracies, they are found in public consumer advocacy bureaus and in non-profit-voluntary health and welfare councils, among others. Thus there is a complete continuum of roles, actors, and sponsoring organizations.

As individual actors we can be the passive receptors of change, or active participants in the process. What then, we ask, motivates participation in change? Schein suggests three factors which stimulate role innovation. First, the world continually changes, generating new problems to be solved which are legitimate to the actor. Even the most conservative medical practitioner may be genuinely excited by a more effective treatment regimen for a serious medical problem. The Jewish Neighborhood House which discovers that its starting basketball team is all black is made vividly aware of its rapidly changing neighborhood. The senator or reporter who is mugged discovers a new world of assaultive behavior. The plight of the aging, sick parent adds a new dimension to the life of the bank president. Schein calls this "role suction". All individuals, conservative or liberal, may find themselves dragged into new situations and directions as society directly confronts them.

A second reason for engaging in a change agent role is that individuals tend to redefine their jobs in a way most congruent with their interests and style. It can be to limit. It can be to expand. If the latter, it enhances the possibility of invention.

Thirdly, deliberate, logical, rational efforts can be made to stimulate role flexibility or role change. Undergraduate and graduate schools may participate; continuing education may be designed. Not only are there town/gown struggles relative to discontinuous values; but, problems such as a medical school in conflict with the local medical society regarding the teaching program and the role of the faculty in the community may occur. The predictable, comfortable, status quo is the customary choice, and resistance to this change strategy can be expected.

THE LOBBYIST

Another very popular American game is lobbying. Sanctioned and legitimized lobbying is surrounded by laws, rules, regulations, and rituals. And according to the critics of the practice, the existing strictures are both commonly abused, and inadequate. Fundamentally, the goal of the game is to enhance one's interest, and, failing that, to protect that interest. Your interest may be personal, organizational, or in the nature of an extended network. It may be self-centered or humanitarian.

The size and cost of the lobbying effort is unknown but it, obviously, is substantial. According to Consumers Union the grocery industry alone has more than 100 specialized national trade associations! The Pharmaceutical Manufacturers Association in 1972 employed 70 people on a three to six million-dollar budget (Consumer Reports, 1972).

But lobbying need not be exclusive to the business sector. Elected officials go before the public every two, four, or six years. As such, they must be responsive to a complex, pluralistic, often-in-conflict constituency. The American Medical Association, it is claimed, on issues highly significant to its interests, contacts the family physicians of the various congressmen. Consider sermons or family visits to elected officials by the clergy. Chaplains to the legislature are well known for their political skills. Those thousands of welfare mothers were loudly, and clearly heard as they flocked to their representatives' offices.

The rules of lobbying are not a deep, dark secret. However, the choice of some tactics (e.g., sit-ins) does not enhance the use of other change means (e.g., consensus or favors). David Kinzer, the lobbyist of the Illinois Hospital Association, suggests some do's and don't's. Tell the truth, and listen. Find out what the elected official is saying. Determine who among your constituency knows the legislator (especially those who have made substantial campaign contributions). Know your legislators. Invite them to lunch. Meet with them. Learn to be political. (Kinzer notes that hospitals are the third largest business in Illinois and that elected officials now make many of the important decisions affecting the hospitals). Develop some credentials with the legislators (in Massachusetts where the general legislative attitude toward organized medicine is negative, it is usually wise to separate the interests of hospitals from those of the medical society). And finally, embrace the youth movement (Kinzer, 1972).

THE OMBUDSMAN

A relatively new actor on the policy influence scene is the ombudsman, defined by the American Bar Association as:

> an independent government official who receives complaints against government agencies and officials from aggrieved persons, who investigates, and who, if the complaints are justified, makes recommendation to remedy the complaints. (ABA, 1969)

This definition is typical of that found in the literature; however, it is incomplete in two respects. Increasingly, ombudsmen are being found in private organizations in addition to government; large non-profit universities are one recent sponsor. The other tenuous point in the definition has to do with the degree of "independence" of the ombudsman. As noted later, there are many exceptions.

The concern of an ombudsman in an organization is principally with the information input-throughput system. By expanding and legitimizing the channels through which information can be received, he can contribute to opening the system and identifying organizational problems which can emanate from policy, problem, or individual issues.

Ombudsmen can operate at a national or lesser governmental level of jurisdiction, or as members of specific organizations. In a 1970 review of national ombudsmen officials in nine countries (not including the United States where the office does not exist nationally), Bernard Frank (1970) found certain common trends; the ombudsman:

1. Is usually the arm of the legislature.
2. Nevertheless, is generally independent of the legislature.
3. Is usually influential and prestigious depending on effectiveness.
4. Typically receives complaints directly about governmental agencies.
5. Conducts an independent investigation.
6. Uses informal, inexpensive and rapid procedures.
7. Cannot give orders or impose sanctions, except in Sweden and Finland.
8. Issues frequent reports on investigations including reasons for dismissal of complaints.
9. Has the authority to inspect agencies.
10. Is readily accessible, and without charge, to the complainant.
11. Usually may suggest administrative or legislative remedies.

A similar review was conducted by Paul Dolan (1969) of ombudsmen-complaint-handling services in nine United States cities. Similarities and differences from the nine country survey should be noted. Typically, the chief executive officer of the city appointed the ombudsman. In most of the cities the complaint body also served as information and general service agency. Reporting and record keeping was generally inadequate. Ordinarily the director was not a well known and respected figure. Staff was limited in numbers. Complaint procedures were increasingly formalized. Thus, on many critical points the American experience compares adversely with the non-American national ombudsman described earlier.

Common to both the European and American experiences, ombudsmen receive complaints and investigate them, and they lack formal power over the administrative branch. The formal complaint-handling mechanism has not transplanted well in America and has developed only slowly in other countries; for it was first established in Sweden in 1809 and not adopted elsewhere until Finland began it in 1919 (Frank, 1970). Perhaps this delay has had to do with our level of expectations and the increasing complexity of our institutions. But, the need is apparent. The extraordinary range of complex organizations and the growth of government around the world requires some means of dealing with injustices, whatever the source.

The ombudsman is one of many alternative ways of dealing with the bureaucratic jungle. Each of many forms of intervention has its value, and each may play multiple roles. The courts and legislative bodies serve to correct system failures and individual bureaucratic transgressions. Yet, as the legislative bodies move intensively on matters of individual justice their ultimate responsibility to act on larger public policy issues is hampered. The courts are increasingly constrained as a backlog of cases emphasizes their inability to respond. Further, many human problems with bureaucracy may stem from simple clerical errors, laggard officials, or maladministration.

Many of the problems which most aggravate the public are not matters for litigation or legislation. Historically, most problems have been handled by the offending agency itself, as well they should. Outside advocates of the type enumerated elsewhere in this paper play additional roles. And, other interesting advocacy developments are those of the organized media, ranging from investigative reporting to "action lines". And of course, muckraking, popularized at the turn of the century, continues to serve valuable functions today. Despite these and many other alternative intervention mechanisms, the formal grievance-complaint procedure has a role. The problem, as with all mechanisms, is to maximize its potential, minimize its weaknesses, and to recognize that it is only one part of the grand non-system.

Problems emanate from a series of sources. They may be individually-based reflecting the "little old lady" on the corner or agency directors who have been doing it their own way for many years, while the world and the agencies change around them. If this be the case the source of the problem should be identified and dealt with on both a generalized and specific basis. Often in this type of situation the problem of an individual complainant may be resolved satisfactorily but with no change in the continuing behavior of the offending actor. Thus the infraction is continued over time.

Illustrations abound. The Agency Budget Director interprets law and regulations in a way which not only inhibits the agency but is actually contrary to the practice of most other comparable agencies. The agency, its staff and clients all suffer. For example, a public official refuses to credit the European gymnasium as equivalent to the first two years of American college, thus affecting the status of European-trained professionals. The explanation — it would be un-American to suggest that the gymnasium is superior to the American high school.

On many occasions sheer rudeness, arbitrariness, arrogance and inefficiency are at work. Retraining, better supervision, clearer regulations, demotion or discharge are available remedies. Again the issue is an individual one, and individual remedies need to be taken.

It is when broad scale maladministration, or unfair, or improper policy is at work that intervention must be more sophisticated. And it is at this point that the generalist ombudsman is weakest. The ombudsman can identify the manifest problems, but translating these into systemic or major organizational changes may require a programmatic or agency sophistication beyond his ken. The quality of his public reports is of great importance here, for the administrative or legislative branches of government or outside advocacy organization may have to enter into the change arena in order that his recommendations be carried out.

The potential for policy influence by ombudsmen thus depends on their ability to aggregate and analyze cases so that the causal elements can be segregated from those which are the result of inadequate or inappropriate policy. It requires a policy sense. It also requires effective avenues of communication in order for policy makers to be aware of the policy problems. Third, it requires an ability to design structural and systemic procedures to bring the policy in line. And finally, it requires an ability to modify the offending policy.

CITIZEN PARTICIPATION

Advisory committees, trustees, boards of directors, and commissions are some samples of the organized ways in which people participate in, and influence, their public and private organizations. They may also serve as individual volunteers in hospitals, schools, and social agencies, playing highly significant roles in client care. And as individuals they may act to correct specific injustices. However, although important to the client served, this volunteer role is not designed for significant organized policy input and thus will not be discussed further.

A tradition in the health system, the trustee role, is represented by the hospital or health agency board of directors. The hospital model is the most classic, and since hospital costs represent more than 40 percent of all health costs we will focus on it as an example. The non-profit voluntary general hospital, traditionally financed principally by the fees of individual users and philanthropic donations, has been governed by large givers — those financial influentials in the community who could most easily help to reduce the hospital's annual operating deficit. Involved in policy making and problem solving, they become personally invested in the institution and work for its excellence, continuance, and financial stability.

It is also suggested that hospitals are essentially business corporations differentiated primarily by their not-for-profit status. Eppert (1968) notes that businesses and hospitals both must plan, do research, schedule legal competency, cope with financial problems (debts, receivables, payables, cash flow, depreciation, maintenance and capital expenditures) and develop marketing skills. There are differences of course; dividends in hospitals go to the patients in better health care. There is no stockholder equity and there is no profit or loss statement as a performance measure. If the trustees' function is perceived as fund raising it is clear that the capacity to influence significant givers is the essential competence. If their function is perceived as policy making, then entree to the wealthy community of givers is not required, as policy is neither a technology nor the property of any economic class. If the board is also invested in organizational problem solving (e.g., finance, research, personnel, etc.) then certain management skills would be required. Although the latter are not necessarily a function of the upper classes, they do reflect certain professional technologies ordinarily found in the upper-middle and lower-upper classes.

Thus, the determination of function significantly influences the types of people chosen to serve as board members of hospitals and other human service institutions. In turn, this helps to better understand the problems of the community and consumer control and participation debates of the 1960's — rhetoric or reality. To reiterate, if the trustees' principal function is to raise money then they cannot be representative of the community, for money and influence are not equally distributed throughout the community.

But hospitals are no longer principally financed via philanthropic donations. Third party arrangements now dominate hospital financing and as prepaid procedures are expanded, especially if a national health insurance plan is adopted, the requirement of financially elite boards will be increasingly less significant.

Another role requirement suggested above is the management-consultant role and his participation in the implementation of intra-organizational administrative policies. The lawyer board members contribute their time as individuals, or on standing or ad hoc committees on legal issues. The bankers and investment counselors advise on financial matters. The trustee head of the large accounting firm participates in the development of more effective cost control systems. The net result is that highly competent people giving of their skills and time enhance the organization's competence and reduce what otherwise would be direct costs to the hospital. But it can be determined that need for effective public policy dictates that representativeness on boards is more important than the free skills of the professional and managerial elite. Costs may increase as a result, but as hospitals are enlarged in size and complexity the capacity to effectively capitalize on the skills of volunteer trustees is likely to be reduced anyway.

We are suggesting that there have been very real reasons for hospital and similar boards to be dominated by the financial, social, and technical elite. A Greater Detroit study in 1971 found that hospitals were:

> dominated by business executives, members of the legal and accounting professions, and spokesmen for medicine and hospitals. Representatives of the consumer and the general community are very seriously under represented. (Goldberg and Hemmelgarn, 1971)

Underlying this discussion is one of role and function. The trustees should be maximally functional in time and space. If hospitals are changing, trustees should reflect the changing needs. One can reasonably predict that voluntary not-for-profit general hospitals will be required to make increasing organizational changes in succeeding decades. Many will be eliminated as too small or inefficient. Others will be incorporated into prepaid group plans. Some will eliminate certain services and share with other institutions. Others will merge. Financial incentives and controls will increase. The most significant policy decisions will increasingly be made at a distance from the given hospital, and most will be public decisions. They will be required to cooperate with others. The focus will be on preventing hospitalization, or on earlier discharge. Alternatives will be expanded. Both health care delivery and management will become increasingly technical and complex. If these and other trends reasonably reflect the likely changes, should governing boards of volunteers be continued, and if so, what roles should they play?

I suggest that as hospitals and other human service voluntary associations are increasingly bureaucratized (the inevitable consequence of the above predictions) citizen boards will become more, not less, important, but that their roles will change. They will still raise money but its proportion of the annual operating budget will continue to shrink thus reducing the importance of this function. They will continue to play technical advisory roles but increasingly these technological needs will be filled by full-time staff or purchased as necessary from other organizations. The trustee policy roles will be delimited by governmental and third-party statutes and regulations. Contrariwise their input role will gain importance as they are increasingly required to serve as a bridge from the community and consumer to the health care and management technocrats. Their principal function will be to act as an early warning system to prevent the institutions from succumbing to the natural organizational tendency towards inflexibility and dehumanization.

The Holy Grail of the 1960's Office of Economic Opportunity was the "maximum feasible participation" of the poor in the policy and decision-making of local community action agencies. These were significant experiences and their implication to the human service industry will be discussed now based upon the hospital trustee as an ideal type. Hope/hustle, real/unreal, opportunity/snare are all judgments made, at one time or another, of the maximum feasible participation policy. In our discussion of hospital trustees we have suggested that ideology must be separated from reality and that a pragmatic analysis of functions related to time is the relevant analytic frame.

One point of confusion has to do with the equation of poor with consumer, with community, with citizen, and the equation of each separately and collectively with the common good. They are often not the same, but the confusion is significant for it suggests some larger issues at work. Hershey (1970) suggests "America must face the challenge of increasing urbanization and societal scale, exploding technological development and continued growth of bureaucracy if it is to remain a free and democratic society." The terms alienation, estrangement, and distance have all been used many times by many authors, and although they do reflect a broad gestalt it is possible to sharpen the issues within the frame of citizen input. One critical set of variables, essentially non-class, non-ethnic, non-sex and non-race related, are the problems created by the increasing scope and complexity of modern science. Some of our young and not-so-young people are raising serious questions about the effects of technology and its unbridled influence. Carroll (1971) and Mesthene (1969) have described it as "the anti-technology spirit that is abroad in the land." Complementary to these developments and concerns is the growing realization that our social goals are increasingly technologically dependent. To Marcuse (1964) ". . . what is at stake is the redefinition of values in technical terms, as elements in the technological process." Carroll (1971) describes participatory technology as one of the countervailing strategies to make science and technology more responsive. Essentially, it is a development which brings people into control over the agents of technology. This development, as with others described in this paper, also implies increasing dissatisfaction with the workings of representative democracy at the elected and bureaucratic levels. The fact that the solutions create new problems will be discussed later.

In addition to the technical vs. non-technical dichotomy, the current participation-control ethic can also be cast in other terms: racial, ethnic, social class, local autonomy vs. larger community, citizen vs. bureaucracy, civil servants vs. elected officials and consumer vs. provider; it is clear that this analysis only represents a single facet of a complex array (just as any of the castings above over-simplify). This is not to suggest that all of the issues are found in each episode, but rather that it is unlikely that any can ever be exclusive.

What, in fact, is meant by these magical catch-phrases?

> In its ideal form, participation consists of sharing actual decision-making power among politicians, public administrators, and lay citizens in planning, programming, and administering policy. (Hershey, 1970)

Citizen participants in the above model would be representative of all. However, another model is more community based, implying some sort of geographic catchmenting, ordinarily a neighborhood or cluster of neighborhoods. The residents of the area are then represented. Alternatively, one could draw from the ranks of those served, the users or consumers, past, present, or potential. None of the models imply any positive or negative loading on class, racial, or ethnic grounds. It depends upon the service offered, its location, and its clientele.

A public welfare program is clearly delimited to those below a certain income level. The taxes supporting the program are drawn from higher income levels. An urban Boys Club in a Black neighborhood is usually limited by the age of its users to a one-mile geographic radius. Thus its users will generally reflect the immediate neighborhood, partly modified by the availability and location of public transportation. A large complex urban teaching hospital will undoubtedly serve its neighbors, especially in its outpatient clinics and emergency wards, but may also receive world-wide referrals depending on its reputation and specialized skills.

Requiring consumer representation on a health planning council merely means that they should not derive their principal income from the health industry, for all people including providers, are health industry consumers. If it were also required to be representative of its geographic area then, depending on definitions, we could add such variables as sex, race, ethnicity, age, class, religion, etc. It is also clear that there are no effective means for populating boards and committees consistently and effectively from the lower-lower (or poor-poor) classes.

What then are the goals? Essentially they are the converse of needs. People are unhappy about their bureaucracies and efforts are being made to accommodate this dissatisfaction. One method of reducing this problem is to involve them in policy and decision-making. Those feeling excluded want to live and feel like full and equal citizens. They want an opportunity to participate in their own destiny. Thus, we have both process and product goals. They want participation at all levels and they want results. They want system reform.

Policy and advisory boards and committees are not new. Participation is not new. Representation is not new and elections are not new — nor is the use of conflict to modify power. The American organized labor movement successfully fought this same battle in the 1930's and has been over-vigilant since. Woe be the public or private administrator who excludes organized labor from his volunteer committees, whatever their functions. The value of this continuing participation by organized labor is clearly evident and no one would seriously challenge it now. Similarly, general purpose local government, state government and the Federal government have long involved the citizenry, often on a specialized basis. The U.S. Department of Agriculture has included local farmers in its extension service planning for decades. The U.S.

Department of Health, Education, and Welfare has a similar tradition in the utilization of technically skilled people as advisors to its research and training grant program. At the policy advisory level, committees extending back for more than a century can be found. Although the New England town meeting model is the ideal toward which many aim, these same small towns have always had many small standing and ad hoc advisory committees in addition to the traditionally elected bodies.

The voluntary human services field is exemplary in its utilization of citizen volunteers at many operating levels. Thus the issue is not whether to use or not use committees, but rather with how much power, and how and from whom they are to be selected. The Board and Committee room is an important battleground, many decisions are or can be made there.

What then are the problems which can be anticipated? The classic area of controversy is that of power. Hochbaum (1969) envisions each group — the consumers and professionals — as possessed of decisional territories. Placed together there are two overlapping circles. One circle includes the issues which the consumers claim for themselves, the other contains the professional prerogatives. The overlapping area represents the potential conflicting turf. Ultimately the field of difference can be narrowed if the health professional is willing to limit his prerogative to technical issues; although, even with this more explicit and limited role many differences are still possible. For example, is national health insurance a general social policy issue, or one requiring medical competence to make a judgement? Does a State Commissioner of Mental Health have to be a physician? Who decides that which is technical and that which is not?

Another set of potential conflicts revolves around wants, needs and equity. Obviously only the people themselves can determine their wants but who determines needs and equity? Norman Lourie (1967) casts the problem starkly — what if each neighborhood wanted to establish its own sewerage system? Essentially these and other differences can be dealt with as political issues to be negotiated, or as technical-rational issues, however, Carroll (1971) suggests a number of potential consequences resulting from these strategies. Politicalization can deteriorate into obstruction, widespread veto and paralysis of public action. Contrariwise, the skilled administrator may manipulate the participants to generate the illusion of public activity and support to achieve ends which may be inimical to the greater good. One evident result of some of the power tactics of the 1960's was a clear system-overloading, creating serious organizational goal displacement. Also possible is the shifting of influence from one small group to another still leading to the tyranny of small decisions.

Participation by those with low income further compounds the problems. The poor have had few resources and few experiences with the subtleties of organized board decision-making processes. To function maximally effectively at the policy level boards must be future oriented. But when income is seriously limited, survival today is the agenda. Chilman (1966) suggests that poverty reinforces a physical rather than verbal response; authoritarian rather than democratic interactive styles. Harry Specht (1966) adds additional barriers: self-defeating attitudes by the poor themselves reinforced by a negative community attitude toward them, organizational practices which discourage their participation, and the general reluctance of those in control to admit new participants. O'Donnell and Chilman (1969) remind us that the critical task is to maintain participation over time, and that support and positive reinforcement is necessary.

The method of selection is also of concern. Who are these people making policy and from whence do they come? We have already referred extensively in this section to the composition of boards and committees, using the general hospital as an analytic model. We concluded earlier that boards should be designed relative to their functions. We have also stated that some sacrifice in technical competence may be justified by gains in broadened representation. But how representative is representation and how necessary? Even at the New England town meeting some people chose not to attend. Some attend but do not participate. The contemporary procedure available to truly and accurately reflect public opinion is the household survey by skilled survey researchers. Election is another model, but the rate of participation in OEO and Model City elections has seldom exceeded five percent of those eligible. Further, the good government reformers and specialists in public administration have long advocated the elimination of the long ballot. Their argument is that the ballot should be as brief as possible in order to persuade people to attend to the significant issues. The longer the ballot the less discriminating and objective are the voters. They advocate fewer elected and more appointive officials — a model quite contrary to the maximum feasible participation ideology. But both are considered "white hat" reformers.

Other requirements contributory to meaningful representation as a goal are the simple requirements for all policy makers — they need certain kinds of analytic and verbal skills, and it is usually the motivated people with these skills who become visible and active. A traditional finding of research on boards and committees is that even those groups have deliberately broadened their ethnic, racial and social class base and have consistently selected leaders from those people who are motivated and verbal. But leaders, by definition, are not representative. Yet, what is wrong with natural leaders? Very likely, appointments, not elections, will remain the operating procedure for securing board members; most boards will move toward elaborating and extending their base of membership, and those chosen will have already demonstrated certain leadership qualities.

CONCLUSIONS

The subtle and difficult task for change seekers is the blending of goals and objectives; the reduction of alienation; the establishment of a more open, responsive, efficient, available, accessible, continuous system of high quality and reasonable cost. It is quite appropriate that objectives occasionally will be in conflict with each other, but excessive discontinuity can be seriously dysfunctional. The consumer may want a health center in every neighborhood, but based upon analysis of need and maximization of resources a client group of at least 30,000 per health center may be required. How are these and similar questions to be decided? The goal of availability may have to be balanced by the goals of quality and cost effectiveness. These are policy questions requiring technical information so that policy makers can respond in a rational manner.

Remember the underlying premise: Change is inevitable; change is continuing; change is occurring at an accelerating rate; and policy change is the most significant form of change. The means are many but not always complementary. Four change processes have been described in this paper. Others were noted earlier in the introduction. The choice depends upon a number of variables: sponsor, sanctions, skills, opportunity, and resources. The target-policy, may be public or private. In either case it is usually fluid and flexible, constantly in tension and change. Change will usually be incremental and may not always be positive but its very existence is promising. The responsibility to impact is that of each citizen. An invitation need not be awaited.

REFERENCES

Alinsky, S. *Rules for Radicals.* New York: Random House, 1971.

American Bar Association. "The Ombudsman." *American Bar Association Journal,* Administrative Law Section, Ombudsman Committee, House of Delegates, 1969, p.1.

The American Heritage Dictionary of the English Language. Boston: The American Heritage Publishing Co., Inc., and Houghton Mifflin Co., 1969.

Carroll, J. D. "Participatory Technology", *Science*, 171:2972, 1971, 647-653.

Chilman, C. S. "Growing Up Poor." Welfare Administration Publication No. 13, U.S. Department of Health, Education, and Welfare, May, 1966.

Consumer Reports. "The Unmaking of a Consumer Advocate." 37:2, 1972, 80-83.

Department of Health, Education, and Welfare, *Towards a Comprehensive Health Policy for the 1970's.* A White Paper, May, 1971.

Dolan, P. "Pseudo-Ombudsman." *National Civic Review,* VLIII:7, 1969, 297-301.

Eppert, R. R. "Place of the Governing Board in Hospital Management." *Hospitals,* 42:23, 1968, 36-40.

Fischer, J. "Seattle's Modern Day Vigilantes." *Harper's Magazine*, 238:1428, 1969.

Frank, B. "The Ombudsman and Human Rights." *Administrative Law Review,* 22:3, 1970, 467-492.

Goldberg, T., & Hemmelgarn, R. "Who Governs Hospitals?" *Hospitals*, 45:Part I, 1971, 72-79.

Harper's Magazine. "Seattle's Modern Day Vigilantes." 238:1428, 1969.

Hershey, G. "Strategies for Change." *National Civic Review,* LIV:1, 1970, 15-20.

Hochbaum, G. M. "Consumer Participation in Health Planning: Towards Conceptual Clarification." *American Journal of Public Health,* 59:9, 1969, 1698-1705.

Kinzer, D. M. "Extracts from 'Confessions of A Lobbyist'." *Trustee,* 25:2, 1972, 26-31.

Lourie, N. V. "Orthopsychiatry and Education." *American Journal of Orthopsychiatry,* 37:5, 1967, 836-842.

Marcuse, H. *One Dimensional Man.* Boston, 1964, p. 232.

Mesthene, E. *Technology Assessment.* Hearings before the Subcommittee on Science, Research and Development of the House Committee on Science and Astronautics, 91st Congress, 1st session, Government Printing Office, Washington, D. C., 1969, 246.

The National Association of Social Workers, *Goals of Public Social Policy,* New York, N.Y., 1959.

O'Donnell, E. J., & Chilman, C. S. "Poor people on public welfare boards and committees — Participation in policy making." *Welfare in Review,* May-June, 1969, 1-10.

Sanders, I. T. "The Involvement of Health Professionals and Local Officials in Fluoridation Controversies." *American Journal of Public Health,* 52:8, 1969, 1274-1287.

Schein, E. H. "The Role Innovator and His Education." *Technology Review*, 72, 1970.

Schorr, A. L., & Baumheier, E. C. "Social Policy." *Encyclopedia of Social Work*. Robert Morris, Editor-in-Chief, National Association of Social Workers, New York, 1971, 1361-1376.

Specht, H. *Urban Community Development: A Social Work Process*. Centra Costa Community Services, 1966, 27-28.

Time Magazine. "The U.S.'s Toughest Customer." December 12, 1969, 89-98.

Time Magazine. "Radical Saul Alinsky: Prophet of Power." March 2, 1970, 56-57.

Time Magazine. "Breaking the American Stereotypes." February 14, 1972, 36-42.

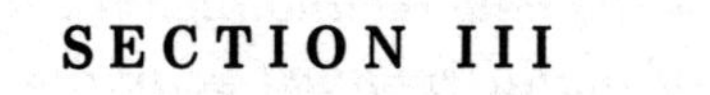

SECTION III

Behavior Change and Control Technologies

No matter how well we organize treatment programs and delivery systems, our overall effectiveness will be strongly related to the quality of our intervention technologies. At some point in the intervention process, reasonable change strategies must be implemented in order to create the crucial outcome of client behavior change. It is at this point that the new technologies generated by behavior modifiers must be put into effective practice.

In the following section, the behavioral model of treatment is presented in overview and reflected in treatment programs designed for specific client populations. Nothing is wrong with a concept of illness or disease that is under the skin of an individual and can be cured by medical or surgical intervention. However, the long-standing debate over the appropriateness of applying the disease or medical model to psychiatric patients in whom no biological pathology can be identified continues to rage. In contrast, the behavioral model insists that behavior is the treatment target rather than a symptom of some other problem which should be the object of treatment activities. This model rules out "disease" as a legitimatizing concept and, consequently, forces practitioners to ask what is ethically permissible intervention.

As Ullmann and Kemp correctly note in their chapter, "Home Intervention Training Programs", an emphasis on delivery without change of model frequently leads to the creation of junior psychiatrists and the traditional position that it is who *provides the service, not* what *is delivered, that is the crucial point. The model which is used to generate intervention strategies has significant implications for the settings in which treatment takes place, who provides the treatment, and the type of intervention considered appropriate. The behavioral model, in contrast to psychodynamically-oriented, traditional models, leads to improvements in intervention outcome and helps to solve many of the practical problems related to the issue of finding sufficient manpower to provide adequate treatment.*

For example, use of the behavioral model allows the mental health professional to maximize the role of nonprofessional change agents. As Ullmann and Kemp point out, effective use can be made of parents and teachers as the primary "therapists" for problem children. Their chapter outlines a concrete training program designed to teach mental health workers how to help parents and teachers change problem behaviors in the home and the classroom.

Kazdin, in his chapter, "A Review of Token Economy Treatment Modalities", shows how the behavioral model leads to a greater emphasis being placed on the work of non-traditional mental health workers, who are the critical links in both the hospital treatment of chronic patients and their ultimate maintenance in the community. Treatment generalization and response maintenance in the natural environment demand the therapeutic utilization of many people other than professional mental health workers.

Perhaps behavior modifiers have had their greatest impact on client populations that no one else wanted to treat. Chronic mental patients are notoriously refractory to treatment and, thus, the newer behavior therapies have been allowed to attempt behavior change in this population. Kazdin's chapter is an excellent overview of the recent advances and emerging trends of token economy treatment systems for institutionalized psychiatric patients. Implications related to the changing nature of treatment, the organization of treatment staff, and the development of adequate community support systems are discussed in this chapter, along with the issues of the long-term maintenance of behavior changes and the transfer of learned responses to different settings. A convincing case is made for the efficiency of behavioral treatment programs and the importance of a system's orientation in the establishment and continuance of them.

Atthowe's chapter, "Behavioral Innovation: An All-Encompassing System of Intervention", stresses that mental illness and mental health need to be viewed in the broader context of everyday living, if we are to be more successful with our treatment systems. The ultimate human service delivery system would be a good deal more all-encompassing by including maintenance and prevention as well as treatment. This suggestion by Atthowe is in response, partially, to the many studies which demonstrate that treatment gains, no matter how successful, are very shortlived because the extra-treatment environment does not support the gains.

Atthowe also provides some guidelines and practical suggestions for developing, utilizing, and evaluating comprehensive systems of intervention.

Radically new treatment systems, based on the behavioral model and, perhaps, not located in any of the traditional mental health agencies, may prove to be the answer to most of the problems in living which are so costly to both society and individuals. New roles for behavior modification specialists are going to be needed, especially if it is true that the contingencies controlling a professional's behavior just do not allow for the type of radical new approaches which seem necessary. A data base for administrative and therapeutic decision-making is desperately needed and the behavioral model offers some hope in the development of accountable systems which can support necessary action and the reorganization of flexible resources.

Leonard P. Ullmann

Carolyn H. Kemp

In the chapter that follows, Ullmann and Kemp articulate a behavioral model of intervention and deal with the implications of the model for treatment. They then specify a home intervention program which exemplifies these considerations.

Ullmann and Kemp point out the importance of the realization that there are genuine differences between the behavioral model of intervention and other models, especially the medical model. By ruling out "disease" as a legitimizing concept, behaviorists are forced to deal with the ethics of what is permissible intervention, and to recognize that the act of labeling people may have extremely negative social consequences. In addition, diagnostic labeling provides little in the way of useful treatment information.

The behavioral model of intervention calls attention to the fact that under many conditions nonprofessional intervenors are more effective than traditional therapists in producing significant client changes. In this chapter, parents and teachers are the focus of training programs designed to provide them with the skills needed to manage problem behaviors of children both in home and school settings. Careful scrutiny of the training programs is essential, according to Ullmann and Kemp, who explain the behavior of the intervenors by using the same model and concepts that the intervenors are being trained to employ. This chapter's focus upon observable behavior and clear outcome evaluations of training programs brings us closer to accountable intervention systems.

CHAPTER 8

Home Intervention Training Programs

This book deals both with general conceptual approaches and the applications stemming from them. It examines both philosophies and their impact. The present chapter[1] is divided into a sketch of a model and one set of applications and solutions that follow from it. It articulates the chapters by Atthowe, Kazdin, and Malott, by discussing some specifics of *what* service is delivered, and thus augments the earlier papers on the form of the organization that provides services. That is, we move from process to content and goals.

THE BEHAVIORAL MODEL

The view reflected in this paper may be called sociopsychological or educational. It has been presented in detail in a number of places during the last decade (e.g., Ullmann and Krasner, 1965, 1969) and is similar to papers by Ford and Atthowe which argue against the medical model.

Deviant Behavior

Deviant behavior cannot be treated or controlled unless first identified. But the very concept of abnormal behavior is a human, social enterprise, and the so-called mental illnesses change over time and are diagnosed with low reliability. "Mental disease" is an instrumental act that has consequences — it is an operant.

"Mental health" is a model developed during the last century as a step forward and an alternative to a model of sin and retribution. Both the religious and medical models have served as forms of social control. There is an area of activity in which the individual acts in a manner that someone, including at times the individual himself, finds changeworthy. Such behavior may appear difficult to understand, self-defeating, unusual, or the like, and it may be used to augment and extend written law for purposes of social control.

Intervention Systems

Who does what to whom, when, and for what reasons involve acts that are deduced and legitimized from some paradigm, whether that model be physiological (give pills), demonic (pray like hell), psychoanalytic (talk to an expensive person), client-centered (talk to a less expensive one), humanistic (touch and feel), sociological, or educational. When we talk of new developments, the most important aspects of the behavior change scene are in terms of models rather than specific treatment or person delivering treatment.

The key aspect of a behavioral approach is that action is considered the crucial target rather than a symbol or symptom. Behavior is emitted in situations, and in reaction to them, and with consequences. The focus is not only on actions but on a unit of situation-action-consequence. We teach people how to deal with situations; we do not treat responses alone. We do not treat an abstraction such as schizophrenia, but behaviors that lead to the designation of schizophrenia. The people with whom we work are not schizophrenics, but individuals who react to some, but certainly not all, situations in a way that is worth changing. Even here, whose behavior is changeworthy is subject to social and situational considerations since it is difficult to think of any act that is in and of itself always pathonomic. We do not change mental health, but responses to situations; these may lead to ratings, but we must not confuse scores with what they symbolize.

If no act is in and of itself changeworthy, then the ethics of behavior change and social intervention come under scrutiny. All too often, the "mental health" professional has let himself be used to solve social problems by labeling a person causing difficulty in a way that discredits or even causes him to be incarcerated. In ruling out disease as a legitimizing concept, and in focusing on behavior that is developed and maintained, we must eventually ask what is permissible intervention. In this regard, the behavioral approach differs from many other "community" or "mental health" approaches which do not ask about the appropriateness of intervention.

A new model may have difficulty obtaining funding; at the level of the legislature, the types of social control desired by legislators may not be considered appropriate by the workers who want money to educate and support change agents. There is a problem of public relations; the model needs to be diffused not only to fellow workers, but to the more general public. Further, as time since graduation increases, the professional requires retraining lest one decade's bright people become the next decade's conservatives. In-service and field training techniques need to be developed to supplement more typical academic procedures.

The Intervenors

Availability of personnel, regardless of the model, affects the incidence, if not the prevalence, of behavioral difficulties. There is a danger that a form of Parkinson's law will result: work increases to keep the professional busy. This is not hypothetical: intervention within the medical model makes two presumptions. The first is that of the personality or general within-individual system that must be changed prior to change of overt action. Work on the lack of generality of personality traits, such as that by Mischel (1968) questions this concept. A second related concept is the presumed unchanging or stable nature of psychiatric categorization. Clarizio (1968) and Scheff and Sundstrom (1970) are authors who have pointed out how weak this assumption may be, particularly when dealing with young people.

Approaching the problem from the opposite direction, we may ask about the effectiveness and efficiency of techniques that are currently considered traditional. We should note that there are spontaneous remissions; that is, change without benefit of professional intervention. Such changes have been viewed with the same moral alarm as marital acts without benefit of clergy. But we should face up to the possibility that professional intervention may publicly label a person and may alter the reactions of significant others in a manner that is not only an ineffective waste, but actually a disservice to the client.

There are true medical problems or physiologically based changes in behavior. The neural deterioration of advanced age, tertiary syphilis, and cretinism are examples. Even here, however, training may make a significant difference. At another level, there are conditions of genuine tissue change, such as those due to the use of noxious substances such as tobacco, but ones in which acts by the person are crucial to the difficulty. While tobacco and alcohol at times lead to unpleasant consequences and changeworthy behavior, we must face up to the fact that some intervention decisions are paternalistic. Again, what is "sin", and even if something is sinful, when may we intervene?

The very designation of an area as a social problem, or as suitable for professional intervention, may be a function of sociological or political variables. For example, we may ask what impact the compulsory education and child labor laws had on the development of the concept of a period of childhood. Law influenced psychological concepts. The reverse is also true as in the use of psychological material in the 1954 Supreme Court decision on school desegregation.

Finally, the staff intervenor must be formulated, studied, and trained with reference to the same concepts that are used in the model of the person and behavior that is to be changed.

The Effect of the Model on Intervention

Before detailing our approach to this matter, we would like to provide some background and contrast. An emphasis on delivery without change of model often leads to the creation of junior psychiatrists. In a related manner, quasi-psychoanalytic sessions may be seen in encounter groups and the Maxwell Jones type of community sessions. One extreme of this phase encompasses apostles of warmth, empathy, and genuineness who act as if these words had clear meaning and impact across all situations and people. At times they are used as universal medicines: A mood or process is established and the therapist has little responsibility. Another way of saying this is that there is a concept that a good therapist is a good person and treatment is what the therapist *is* rather than what the therapist *does* differentially and contingently.

Another extreme of the traditional position essentially says that the medium is the message: *who* is so important to the delivery that there is no consideration of *what* is delivered. That a person is an indigenous nonprofessional may or may not help communication, but it should never excuse failure to train that worker in how and what to communicate.

It is interesting how people for whom the traditional professional does not have an adequate service are relegated to the lowest rungs of the clinic's staff. This permits the professional to criticize from a distance. On a larger scale there is the implicit snobbery of giving a minimally trained person the task of delivering a service in the name of economics or because they "talk the language". In the hospital it is the attendant and not the physician who delivers moment to moment service; we are in danger of having community service also delivered by the lowest common worker.

We have mentioned that we are working in an area of social control and we need to ask about the legitimacy of our intervention. We are given a variety of powers and must be alert to the social structures and cultures of our clients. An example comes from child welfare services for Blacks in the United States. "Child welfare services in the United States have focused narrowly on the placement of children away from their families. Few services have developed for children living with their parents, and none have commanded the funds and energy devoted to placement services" (Billingsley and Giovannoni, 1972). To further quote Billingsley and Giovannoni: "Children were to be saved from a life of sin by learning individually to conform to the 'right' standards." Good parents are good people, and bad people don't deserve to keep their children. There is a similarity here between concepts of good therapists and good parents, and between religious and psychoanalytic dogmatism (Billingsley and Giovannoni, 1972).

While not solving all the problems of sociological pressure and legitimacy touched on above, we think the behavioral approach is a step forward. First, the arena of change is the environment in which the target or changeworthy behavior is observed rather than through some intrapsychic focus. The very nature of designating situations, behaviors, and consequences, increases the public nature of the treatment goals, both conceptually, and more importantly, at an evaluative level. The referral itself is translated into a behavior so that we are dealing with specific rather than undefined variables. Such specific changes lend themselves to evaluation and build in a first step for accountability.

Another problem dealt with is that parents and teachers are frequently the people with whom the change agent works. This is a treatment in the home and school that is an alternative to removal to hospital or prison. By avoiding designation in terms of diseases, we may help to protect children from stigmatizing labels.

The goal is behavior that increases satisfaction for *both* child and significant other. For example, effective shaping implies starting where the child is and providing him with response contingent satisfactions. We aim to increase skills that will help the child deal with situations. Such skills need not be conforming. For example, altruism, cooperation, question asking, self-modification, appropriate assertion and creativity have all been successfully increased by direct, behavioral intervention. To the extent that school work is instrumental to later life success, we may be paternalistic in helping a child adjust to school, but the major focus is in changing significant others so that they will make the child's environment more consistent, clear, and satisfying. Finally, we are the child's advocate and often work to make more realistic the levels of aspiration and consequent reinforcing practices of adults.

Some final general points are that we need techniques that may be used with atypical clients, that is, people who are not young, not attractive, not verbal, not college educated, not financially successful, not well-trained in psychological thought and theory, and not faced with existential crises. Another point is that we should strive for a theory and treatment that jibe with each other, with laboratory observations, and with further collection of data.

With these background considerations in mind, we deal with new ideas of the formulation of changeworthy behavior; professional behaviors that may alter such behaviors; and methods for training people in such behaviors. It is not a matter of more of the same, but of meeting needs in a better manner. What follows is one such alternative.

HOME INTERVENTION SERVICES

For the past three years, a training program which exemplifies the educational/behavioral model has been implemented. For lack of a better phrase, the training program has been called "Home Intervention Seminar Series". The term "Home Intervenor" was coined by staff at the Adler Regional Center in Champaign, Illinois, and, much as the trainer wanted to change it, it had caught on so quickly it could not be altered. The training program, or Home Intervention Seminar Series, was designed to teach the participant trainees to help parents to modify or change problem behaviors of children in the home, and to help teachers to manage children's behavior in the classroom. The trainer's target for change was the behavior of the seminar participants, or trainees, whose target for change was the behavior of the parent and/or teacher (or grandmother, boyfriend, etc.), whose target for change was the behavior of the child whose behavior was considered changeworthy in the home and/or in school.

The Background of This Approach

There is a significant body of literature which demonstrates that the behavioral approach is effective with children (Allen et al., 1970; Allison, 1971; Azrin and Lindsley, 1956; Baer, 1962; Baer and Sherman, 1964; Bentler, 1962; Bernal et al., 1968; Breyer, 1971; Coleman, 1970; Hall et al., 1970; Hallsten, 1965; Harris et al., 1964; Kondas, 1967; Lindsley, 1966; McReynold, 1969; Meyerson et al., 1967; O'Leary and Becker, 1967; Patterson, 1961; Peterson, 1967; Rheingold, 1956; Rickard and Saunders, 1971; Stuart, 1970; Thorne et al., 1967; Wahler et al., 1965; Williams, 1959; Wolf et al., 1964). There is also a large body of literature which demonstrates that parents can be taught to use the behavioral approach to change the behavior of their children (Allen and Harris, 1966; Barrett, 1969; Boardman, 1964; Edlund, 1971; Gardner, 1967; Gardner et al., 1968; Hawkins et al., 1966; Johnson et al., 1969; Lal and Lindsley, 1968; Mathis, 1971; O'Leary et al., 1967; Peine, 1969, Ray, 1969; Shah, 1969; Wetzel et al., 1966; Williams, 1959; Zeilberger et al., 1968). There is also a body of literature which indicates that paraprofessional staff can be taught the principles and techniques of behavior modification, and, in turn, can effectively teach them to others (Bushell, et al., 1968; Cowen et al., 1966; Christmas, 1966; Goodman, 1969; Hadley et al., 1970; Hall et al., 1968; Hall et al., 1968; Hart et al., 1968; Holzberg et al., 1967; Klein, 1967; Kreitzer, 1969, in Guerney, 1971; Laws et al., 1971; Lichtenberg, 1956; MacDonald, 1970; Monkman, 1972; Nichtern et al., 1964; O'Leary et al., 1969, Osborne, 1969; Poser, 1966; Rappaport et al., 1971; Reinherz, 1964; Rioch et al., 1965; Thomas et al., 1968; Umbarger et al., 1962; Ward and Baker, 1968; Zimmerman et al., 1969).

The Seminar Series

The Home Intervention Seminar Series was run for two years in Champaign, Illinois, through the Adler Regional Center. The third series (no longer called Home Intervention Seminar Series because of a move which permitted the dropping of that title) was run in Honolulu, Hawaii. In all three series, a group of approximately 25 trainees ranged from professional people (a few with master's degrees and even two with doctorates) to non or paraprofessional people (some with less than high school education). All of the trainees were employed by some mental health or mental health related agency.

The program ran for ten weeks — one full day per week. The training was distributed in this way for several reasons. There is good evidence that distributed versus massed practice is more effective. Also, these people were full-time employees of agencies, and could not afford to spend two full weeks away from their jobs; and in addition, a crucial part of the program was the homework and practice done between each session.

There were only three requirements for admission to the program (aside from a limit of 25 to 30 bodies): 1) trainees must now be, or plan soon to be, working with families who have children with changeworthy behavior; 2) they must pick up a child case at the beginning of the training program and carry it through the program; and 3) they must attend every session.

There were three levels of training going on in this program. The first was the academic or didactic information-giving level. The second was the modeling done by trainers — in fact, in the Hawaii Program, trainers went out on cases with trainees to demonstrate and model the method. The third level was continued case supervision given to trainees by trainers both during and after the training series itself. In fact, in all three of the training programs, the trainees elected to continue to meet periodically and set up further learning experiences for themselves.

The academic content was presented via reading material, lectures, films, discussion, behavior rehearsal, and practice. Reading materials have varied from year to year. In the first series, Whaley and Malott's (1971), *Elementary Principles of Behavior*, an excellent text, was used. However, it proved too difficult for most of the trainees. So, for the second series, the H & H (Hall, 1971) series of five booklets plus a bibliographic card file entitled *Managing Behavior* (Hall, 1971), was used. These were somewhat easier, but still not quite right. The next time, the plan is to try Diebert and Harmon's (1970), *New Tools for Changing Behavior*, and Bernard and Powell's (1972), *Teaching the Mentally Retarded Child*, and see how those go. We also used Una Haynes' (1969), *A Developmental Approach to Casefinding;* and for strictly developmental information, Dr. Spock's (1946), *Baby and Child Care*, was excellent.

Another set of reading materials was provided each trainee. This consisted of xeroxed articles from the literature which were pertinent to the training. The articles were arranged according to the day's session to which they most applied, and put in a ring binder, along with the schedule of sessions at the front. To help make the series more applicable and stimulating to the academically advanced participants, other reading material was suggested and provided; for example, Ullmann and Krasner's (1965), *Case Studies in Behavior Modification*, and the several programmed texts published by Research Press, as well as journal articles specific to their cases or questions.

Included at the end of this chapter is an outline of each day's session, which will give the reader a good idea of content, objectives, process, and materials used in the training program.

There were two major themes running through the ten days of the training program which do not show up in the included curriculum. The first was the use of one's self as a reinforcer for changed behavior on the part of the parent and/or teacher. This is both a rapport building and a shaping process. Second, was the emphasis on the same principles for changing parent/teacher behavior that the trainee wanted the parent or teacher to use to change children's behavior. For example, there was stress on praising the parent/teacher for following the program for changing the children's behavior, and a stress on prompting, rather than negative feedback or confrontation, when the parent or teacher made mistakes or failed to follow the program. Other features were a steady evaluation of programs and a readiness to modify them so that the steps were easier to take; that is, feedback based on data rather than personalities was emphasized.

The trainers also used the principles in changing the trainee's behavior—they shaped, prompted, and reinforced very heavily any changes in trainees' behavior which indicated progress. This, of course, modeled the behaviors they were trying to get trainees to use with parents and teachers. In teaching this particular content, trainers demonstrated through role playing, *how* to talk to mothers and teachers, as well as *what* to say to them. Trainers also talked about what they could say in response to various typical responses by parents and teachers. Their own cases, which the trainees carried throughout the series, provided many live examples for the trainers to use in teaching.

In conclusion, we have tried in this paper to describe a portion of the background, concept, and implementation of the behavioral approach. The approach grew out of a revolt against a quasi-medical model and has implications for the formulation for the reform of people whose behavior is changeworthy as well as for the training of people who help in that change. Above all, however, there must be respect for people and their life-styles. Neither people nor their acts are "sick"; rather they are the result of learning experiences and opportunities. We hope that in our work we provide opportunities that will help people become more skillful in satisfying themselves in ways which also satisfy others.

Daily Program Outline

Day 1

Subject	Content Objectives	Process	Materials
I. Introduction and welcome	I. To explain the concept of Home Intervention A. Definition and philosophical underpinnings	I. Lecture	I. Behavioral interventions
	1. Rationale for use of behavior therapy		1. Literature review on behavior modification's effectiveness with children[2]
	2. Rationale for our training other professionals (diffusions and economy)		2. Literature review on training paraprofessionals[2]
	3. Rationale for training parents as therapists for their own children		3. "Parents as Therapists" (Kemp, 1972)
	4. Rationale for going into the home		4. Oral presentation
II. Explanation of course	II. To explain mechanics of course	II. Lecture	II. See literature on "massed" vs. "distributed" practice

Subject	Content Objectives	Process	Materials
			for learning and other training techniques
	A. Case to carry through		
	B. Required attendance (contingencies)		
	C. Small groups — larger group divided into small groups which remain constant throughout and meet each time	C. Assign to small groups (ideal 5 - 7 members) being sure to break up members of same agency and community, if possible. Skilled, experienced, and knowledgeable leaders essential for productive group discussions	
	D. Introduction of group leaders and information as to where and when they may be reached for case supervision		
	E. Explain Pre-Test/Post-Test and give rationale for administering the tests		
	F. Review of curriculum outline — stressing where meetings are being		F. Mimeographed calendar of events,

Subject	Content Objectives	Process	Materials
	held, what time they begin and end, as well as format (speakers, tapes, group discussion, lectures, films, etc.)		place of meetings, time, etc.
	G. Discussion of materials — how and when they are to be used, cost etc.		G. Xeroxed articles and materials given by Day number, available text(s) and other books, (possible textbooks may be chosen from attached bibliography); suggested — Diebert and Harmon, (1970), the *Behavior Management Series* edited by Hall (1971), plus Haynes (1969), *A Developmental Approach to Casefinding* and Spock (1946), *Baby and Child Care*
III. Pre-Test	III. To provide operant level or baseline or knowledge, terminology, and basic principles of behavior modification	III. Hand out Pre-Test and tell how much time is allotted	III. Mimeographed Pre-Test devised specifically for this group

Subject	Content Objectives	Process	Materials
IV. Basic principles and examples of what behavior modification *looks* like, and demonstration of its effectiveness in variety of settings	IV. To provide first exposure to behavioral principles and techniques — to give a visual impression of what use of this approach looks like — a "hooker" — also a stimulus variety	IV. Film	IV. "Reinforcement Therapy" (SKF, 1969)
V. Pre-Test	V. To use Pre-Test as a teaching tool, by having class grade them (*not good* if you want to measure or evaluate teaching)	V. Hand back Pre-Test, review, item by item, giving correct answer and brief explanation	V. Mimeographed material as above
VI. Basic principles of child development	VI. To teach "typical" developmental behaviors for children of different ages, to provide norm for comparison of referred problem children	VI. Lecture	VI. Spock (1946), mentioned above, and mimeographed material on age, physical growth, developmental steps, and typical developmental problems
	Day 2		
I. Continuation of basic principles of child development	I. See Day 1	I. See Day 1	I. See Day 1
II. Mental retardation	II. To give a modern (current) definition of mental retardation, levels of retardation, explanation of IQ (how it is	II. Lecture	II. Una Haynes, (1969); Frankenburg and Dodds (1969), Denver

Subject	Content Objectives	Process	Materials
	measured, and what it does and does not mean), causes of retardation, prognosis, special education (EMH, TMH, sub-trainable, etc.), indications for insitutionalization, typical behaviors or behavioral deficits, how to assess functional level, and how to work with retardates, their parents, and siblings		Developmental Screening Tool; Functional Level Scale[3]; and *Infant Stimulation and Motor Development*[4]; and other such material
	To show how behavioral principles and techniques may be used with retardates	Film	"Rewards and Reinforcements"
III. Small group meetings	III. To discuss material presented thus far	III. Group discussion	
	To become acquainted with group members — assess level of sophistication and gauge speed of future presentations		
	Day 3		
I. Basic principles and techniques of behavior modification	I. To teach concepts, language, and use of behavioral approach	I. Lecture	I. Mimeographed material such as Whaley and Malott (1971); and Lake (1971)
			A bibliography of behavior modification books and journals

Subject	Content Objectives	Process	Materials
II. Interviewing skills	II. To teach how to talk to parents and what to say to them — a behavioral approach to interviewing. To teach how to teach parents to be behaviorally specific, set goals, carry out a program, and reinforce their child's "good" behavior while extinguishing the "bad" behavior. To teach ways of reinforcing parents	II. Lecture	II. Wahler and Cormier (1971)
	Day 4		
I. Interviewing (continued)	I. To show what the interview as conducted by a behaviorist, with the preceding day's material in mind, might look like To reiterate basic skills for interviewing	I. Video tape, critique and discussion	I. Materials such as Ullmann (1969)
	Day 5		
I. Data collection, interpretation, and reduction	I. To teach what "data" means, why data must be collected, how data can be collected, how to chart data, interpret data, and reduce data	I. Lecture. Transparencies projected by overhead, and video tapes	I. Patterson (1972) and Peterson (1967). Transparencies of various data collection forms; video tapes of behavior which participants view and practice recording from

Subject	Content Objectives	Process	Materials
II. Small group meeting	II. To talk about specific data collection method applicable to each case. To clear up any questions about materials thus far	II. Group discussions	
	Day 6		
I. Small group meetings	I. To go over interviews and data collection with specific reference to each case	I. Group discussion	I. Each group member's initial interview form and data collected
II. Case examples from residential cottages	II. To have persons directly responsible for developing and carrying out a program for a child describe the process by which behaviors were specified, reinforcers determined, and a program written to manage the child's behavior To demonstrate step-by-step process in programming To demonstrate complexities of programming for behavior change	II. Cottage counselors talk about specific child and program	II. Mimeographed materials from actual cottage programs, and published cases such as Hawkins, Peterson, Schweid, and Bijou, (1966)
	Day 7		
I. Small group meetings	I. To talk about programs, problems, and to give program suggestions to individual members	I. Group discussion	I. Tentative programs written by group members
II. Model change plans	II. To teach specific behavior modification techniques of "premacking", modeling, contingency contracting,	II. Lecture, "shaping game", and discussion	II. Examples from literature such as Wahler et al. (1965), Patterson

Subject	Content Objectives	Process	Materials
	in vivo training, shaping, desensitization		et al. (1967), and Stuart (1969)
Day 8			
I. Small group meetings	I. To discuss programs	I. Group discussion	I. Written programs for case of each member
II. Fading out of self and concrete reinforcers, and follow-up	II. To teach how and when to gradually fade out the use of concrete, tangible, or primary reinforcers, if they have been part of the program for modifying the child's behavior. Re-explanation of schedules of reinforcement, and continuous vs. intermittent reinforcement of desired behavior To teach how and when to fade self out of case and terminate To discuss follow-up procedures		
III. Dealing with parents in groups	III. To teach how to utilize self more effectively by having parent groups to pinpoint behaviors, set goals, determine reinforcers and write program for children To teach participants how to teach parents in a group the basic principles and techniques of behavior modification	III. Lecture	III. Mimeographed materials such as Lindsley (1968), and Clement (1971)

Subject	Content Objectives	Process	Materials
Day 9			
I. Small group meetings	I. To discuss material presented thus far, and to work out program problems	I. Group discussion	I. Written programs of each group member
II. Carrying it into the school	II. To teach how to apply the basic principles to a school setting; how to influence school administration and teachers; how and when to program for an entire classroom	II. Lecture	II. Mimeographed material such as O'Leary and Becker (1967), O'Leary and O'Leary (1972), and Staats et al. (1967)
III. Review of entire course	III. To clear up any questions about content presented thus far; to attempt to determine areas that need expansion	III. Discussion and lecture	
Day 10			
I. Post-Test	I. To determine whether academic material was learned since administration of the Pre-Test	I. Hand out same test (or similar form)	I. See Pre-Test under Day 1.
II. Critique of course, discussion and grading of Post-Tests, final questions, discussion of follow-up meetings, collection of case material and presentation of certificates	II. To give trainers feedback on content and process of course. To do preliminary grading of Post-Tests. To answer final questions on content or process or follow-up help To reinforce participants.	II. Discussion	

FOOTNOTES

[1] This work was supported in part by grant number 11938 from the National Institute of Mental Health, United States Public Health Service.

[2] S. W. Eggers, C. H. Kemp and J. A. Kramer. *Development and Evaluation of Home Intervention Training.* Unpublished manuscript, Adler Regional Center, Champaign, Illinois, 1973.

[3] B. Hicks, mimeographed material, Progress School, Decatur, Illinois, 1972.

[4] C. Frichtl and L. Peterson. *Early Infant Stimulation and Motor Development*, Unpublished manuscript, Adler Regional Center, Champaign, Illinois, 1968.

REFERENCES

Allen K. E., & Harris, F. R. "Elimination of a child's excessive scratching by training the mother in reinforcement procedures." *Behaviour Research and Therapy*, 1966, 4, 79-84.

Allen, K. E., Turner, K. D., & Everett, P. M. "A behavior modification classroom for Head Start children with problem behaviors." Experiment I. *Exceptional Children*, 1970, 37, 119-127.

Allison, T. S. & Allison, S. L. "Time-out from reinforcement effect on sibling aggression." *Psychological Record*, 1971, 21, 86-89.

Azrin, N. H. & Lindsley, O. R. "The reinforcement of cooperation between children." *Journal of Abnormal and Social Psychology*, 1956, 52, 110-102.

Baer, D. M. "Laboratory control of thumbsucking by withdrawal and representation of reinforcement." *Journal of the Experimental Analysis of Behavior*, 1962, 5, 525-528.

Baer, D. M. & Sherman, J. A. "Reinforcement control of generalized imitation in young children." *Journal of Experimental Child Psychology*, 1964, 1, 37-49.

Bernard, K. & Powell, M. *Teaching The Mentally Retarded Child.* St. Louis, Missouri: Mosby, 1972.

Barrett, B. H. "Behavior modification in the home: Parents adopt laboratory-developed tactics to bowel-train a 5½-year-old." *Psychotherapy: Theory, Research and Practice*, 1969, 6, 172-176.

Bentler, P. M. "An infant's phobia treated with reciprocal inhibition therapy." *Journal of Child Psychology and Psychiatry*, 1962, 3, 185-189.

Bernal, M. E., Duryce, J. S., Pruett, H. L., & Burns, B. J. "Behavior modification and the brat syndrome." *Journal of Consulting and Clinical Psychology*, 1968, 32, 447-455.

Billingsley, A., & Giovannoni, J. M. *Children of the Storm.* New York: Harcourt, Brace, Jovanovich, 1972.

Boardman, W. K. "Rusty: A brief behavior disorder." *Journal of Consulting Psychology*, 1962, 26, 293-297.

Breyer, N. L., Calchera, D. J., & Cann, C. "Behavioral consulting from a distance." *Psychology in the Schools*, 1971, 8, 172-176.

Bushell, D. Jr., Wrobel, P. A., & Michaelis, M. L. "Applying 'group' contingencies to the classroom study behavior of preschool children." *Journal of Applied Behavior Analysis*, 1968, 1, 55-61.

Christmas, J. J. "Group methods in training and practice: Nonprofessional mental health personnel in a deprived community." *American Journal of Orthopsychiatry*, 1966, 36, 410-419.

Clarizio, H. "Stability of deviant behavior through time." *Mental Hygiene*, 1968, 52, 288-293.

Clement, P. W. "Please, Mother, I'd rather you did it yourself: Training parents to treat their own children." *Journal of School Health*, 1971, February, 65-69.

Coleman, R. "A conditioning technique applicable to elementary school classrooms." *Journal of Applied Behavior Analysis*, 1970, 3, 293-297.

Cowen, E. L., Zax, M., & Laird, J. "A college student volunteer program in the elementary school setting." *Community Mental Health Journal*, 1966, 2, 319-328.

Diebert, A. N., & Harmon, A. J. *New Tools for Changing Behavior*. Champaign, Illinois: Research Press, 1970.

Edlund, C. V. "A reinforcement approach to the elimination of a child's school phobia." *Mental Hygiene*, 1971, 55, 433-436.

Gardner, J. E. "Behavior therapy treatment approach to a psychogenic seizure case." *Journal of Consulting Psychology*, 1967, 31, 209-212.

Gardner, J. E., Pearson, D. T., Bercovici, A. N., & Bricker, D. E. "Measurement, evaluation, and modification of selected social interactions between a schizophrenic child, his parents, and his therapist." *Journal of Consulting and Clinical Psychology*, 1968, 32, 543-549.

Goodman, G. "An experiment with companionship therapy: College students and troubled boys — assumptions, selection, and design." In Guerney, Bernard G., Jr. (Ed.), *Psychotherapeutic Agents: New Roles for Nonprofessionals, Parents and Teachers*. New York: Holt, Rinehart and Winston, 1969.

Hadley, J. M., True, J. E., & Kepes, S. Y. "An experiment in the education of the preprofessional mental health worker: The Purdue program." *Community Mental Health Journal*, 1970, 6, 40-51.

Hall, R. V. (Ed.), *Managing Behavior* (series), Lawrence, Kansas: H & H Enterprises, 1971.

Hall, R. V., Axelrod, S., Tyler, L., Grief, E., Jones, F. C., & Robertson, R. "Modification of behavior problems in the home with a parent as observer and experimenter." Prepublication paper, 1970, Bureau of Child Research, 2021 North Third, Kansas City, Kansas 66101.

Hall, R. V., Lund, D., & Jackson, D. "Effects of teacher attention on study behavior." *Journal of Applied Behavior Analysis*, 1968, 1, 1-12.

Hall, R. V., Panyan, M., Rabon, D. & Broden, M. "Instructing beginning teachers in reinforcement procedures which improve classroom control." *Journal of Applied Behavior Analysis*, 1968, 1, 315-322.

Hallsten, E. A. "Adolescent anorexia nervosa treated by desensitization." *Behaviour Research and Therapy*, 1965, 3, 87-91.

Harris, F. R., Wolf, M. M., & Baer, D. M. "Effects of adult social reinforcement on child behavior." *Young Children*, 1964, 20, 8-17.

Hart, B. M., Reynolds, N. J., Baer, D. M., Brawley, E. R., & Harris, F. R. "Effect of contingent and non-contingent social reinforcement on the cooperative play of a preschool child." *Journal of Applied Behavior Analysis*, 1968, 1, 73-76.

Hawkins, R. P., Peterson, R. F., Schweid, E., & Bijou, S. W. "Behavior therapy in the home: Amelioration of problem parent-child relations with the parent in a therapeutic role." *Journal of Experimental Child Psychology*, 1966, 4, 99-107.

Haynes, U. *A Developmental Approach to Casefinding.* Washington, D. C.: U.S. Government Printing Office, 1969.

Holzberg, J. D., Knapp, R. H., & Turner, J. L. "College students as companions to the mentally ill." In Cowen, E. L., Gardner, E. H., and Zax, M. (Eds.), *Emergent Approaches to Mental Health Problems.* New York: Appleton-Century-Crofts, 1967.

Johnson, S. M. & Brown, R. A. "Producing behavior change in parents of disturbed children." *Journal of Child Psychology and Psychiatry*, 1969, 10, 107-121.

Kemp, C. H. "Parents As Therapists." Paper presented at American Association for Mental Deficiency, Minneapolis, May, 1972.

Klein, W. "The training of human service aides." In Cowen, E. L., Gardner, E. H., and Zax, M. (Eds.), *Emergent Approaches to Mental Health Problems.* New York: Appleton-Century-Crofts, 1967.

Kondas, O. "Reduction of examination anxiety and stage-fright by group desensitization." *Behaviour Research and Therapy*, 1967, 5, 275-281.

Kreitzer, S. F. "College students in a behavior therapy program with hospitalized, emotionally disturbed children." In Guerney, Bernard G. Jr. (Ed.), *Psychotherapeutic Agents: New Roles for Nonprofessionals, Parents and Teachers.* New York: Holt, Rinehart and Winston, 1969.

Lake, A. "How to teach your child good habits." *Redbook*, 1971, 74, 186-188.

Lal, H. & Lindsley, O. R. "Therapy of chronic constipation in a young child by rearranging social contingencies." *Behaviour Research and Therapy*, 1968, 6, 484-485.

Laws, D. R., Brown, R. A., Epstein, J., & Hocking, N. "Reduction of inappropriate social behavior in disturbed children by an untrained paraprofessional therapist." *Behavior Therapy*, 1971, 2, 519-533.

Lichtenberg, B. "On the selection and preparation of the big brother volunteer." *Social Casework*, 1956, 137, 396-400.

Lindsley, O. R. "An experiment with parents handling behavior at home." *Johnstone Bulletin*, 1966, 9, 27-36.

Lindsley, O. R. "Training parents and teachers to precisely manage children's behavior." Paper presented at C. S. Mott Foundation Children's Health Center, March, 1968.

McReynolds, L. V. "Application of time-out from positive reinforcement for increasing the efficiency of speech training." *Journal of Applied Behavior Analysis*, 1969, 2, 199-205.

MacDonald, W. S., Gallimore, R., & MacDonald, G. "Contingency counseling by school personnel: An economical model of intervention." *Journal of Applied Behavior Analysis*, 1970, 3, 175-182.

Marshall, G. R. "Toilet training of an autistic 8-year-old through conditioning therapy: A case report." *Behaviour Research and Therapy*, 1966, 4, 242-245.

Mathis, H. I. "Training a 'disturbed' boy using the mother as therapist: A case study." *Behavior Therapy*, 1971, 2, 233-239.

Meyerson, L., Kerr, N., & Michael, J. L. "Behavior modification in rehabilitation." *Child Development: Readings in Experimental Analysis*. 1967, 214-239.

Mischel, W. *Personality and Assessment*. New York: Wiley, 1968.

Monkman, M. M. *A Milieu Therapy Program for Behaviorally Disturbed Children*. Springfield, Illinois: Charles C. Thomas, 1972.

Nichtern, S., Donahue, G. T., O'Shea, J., Morans, M., Curtis, M., & Brody, C. "A community educational program for the emotionally disturbed child." *American Journal of Orthopsychiatry*, 1964, 34, 705-713.

O'Leary, K. D., Becker, W. C., Evans, M. B., & Saudargas, R. A. "A token reinforcement program in a public school: A replication and systematic analysis." *Journal of Applied Behavior Analysis*, 1969, 2, 3-13.

O'Leary, K. D. & O'Leary, S. G. (Eds.), *Classroom Management*. Elmsford, N.Y.: Peragamon Press, 1972.

O'Leary, K. D., O'Leary, S. & Becker, W. "Modification of a deviant sibling interaction pattern in the home." *Behaviour Research and Therapy*, 1967, 5, 113-120.

Osborne, J. G. "Free-time as a reinforcer in the management of classroom behavior." *Journal of Applied Behavior Analysis*, 1969, 2, 113-118.

Patterson, G. R. "A learning theory approach to the treatment of the school phobic child." Paper presented at the annual convention of the AAMD, 1961, Portland, Ore.

Patterson, G. R. *Families*. Champaign, Illinois: Research Press, 1972.

Patterson, G. R., McNeal, S., Hawkins, N., & Phelps, R. "Reprogramming the social environment." *Child Psychology and Psychiatry*, 1967, 8, 181-195.

Peine, H. A. "Programming in the home." Paper presented at the Annual Meeting of the Rocky Mountain Psychological Association, Albuquerque, N.M., 1969.

Peterson, L. "Operant approach to observation and recording." *Nursing Outlook*, 1967, 15, 3.

Poser, E. G. "The effect of therapist training on group therapeutic outcome." *Journal of Consulting Psychology*, 1966, 30, 283-289.

Rappaport, J., Chinsky, J. M., & Cowen, E. L. *Innovations in Helping Chronic Patients: College Students in a Mental Institution*. New York: Academic Press, 1971.

Ray, R. S. "Parents and teachers as therapeutic agents in behavior modification." Paper presented at the Second Annual Alabama Behavior Modification Institute, Tuscaloosa, Alabama, 1969.

Reinherz, H. "The therapeutic use of student volunteers." *Children*, 1964, 2, 137-142.

Rheingold, H. L. "The modification of social responsiveness in institutional babies." *Monograph of the Society for Research in Child Development*, 1956, 21, No. 63.

Rickard, H. C. & Saunders, T. R. "Control of 'clean-up' behavior in a summer camp." *Behavior Therapy*, 1971, 2, 340-344.

Rioch, M. J., Elkes, C. & Flint, A. A. *National Institute of Mental Health Pilot Project in Training Mental Health Counselors*, Washington, D. C.: U.S. Department of Health, Education, and Welfare, Public Health Service, Publication No. 1254. 1965.

Scheff, T. J., & Sundstrom, E. "The stability of deviant behavior over time: A reassessment." *Journal of Health and Social Behavior*, 1970, 11, 37-43.

Shah, S. A. "Training and utilizing a mother as the therapist for her child." Paper read at E.P.A. meeting, Boston, 1969.

Skinner, B. F. *Science and Human Behavior.* New York: Macmillan, 1953.

Spock, B. *Baby and Child Care.* New York: Pocket Books, 1946.

Staats, A. W., Minke, K. A., Goodwin, W., & Landeen, J. "Cognitive behavior modification: 'Motivated learning' reading treatment with subprofessional therapy-technicians." *Behaviour Research and Therapy*, 1967, 5, 283-299.

Stuart, R. B. "Operant-interpersonal treatment for marital discord." Paper presented at AABT, San Francisco, August, 1968.

Stuart, R. B. "Behavioral contracting within the families of delinquents." Paper presented at the Annual Meeting of the American Psychological Association, Miami Beach, Florida, 1970.

Thomas, D. R., Becker, W. C., & Armstrong, M. "Production and elimination of disruptive classroom behavior by systematically varying teacher's behavior." *Journal of Applied Behavior Analysis*, 1968, 1, 35-45.

Thorne, G. L., Tharp, R. G., and Wetzel, R. J. "Behavior modification techniques: New tools for probation officers." *Federal Probation*, 1967, 31, 21-27.

Ullmann, L. P. "Making use of modeling in the therapeutic interview." In R. D. Rubin and C. M. Franks (Eds.), *Advances in Behavior Therapy*, 1968. New York: Academic Press, 1969, 175-182.

Ullmann, L. P., & Krasner, L. (Eds.) *Case Studies in Behavior Modification.* New York: Holt, Rinehart, and Winston, 1965.

Ullmann, L. P., & Krasner, L. *A Psychological Approach to Abnormal Behavior.* Englewood Cliffs, N. J.: Prentice-Hall, 1969.

Umbarger, C. C., Dalsimer, J. S., Morrison, A. P., & Breggin, P. R. *College Students in a Mental Hospital*, New York: Grune and Stratton, 1962.

Wahler, R. G., & Cormier, W. H. "The ecological interview: A first step in outpatient child behavior therapy." *Journal of Behavior Therapy and Experimental Psychiatry*, 1970, 1, 279-289.

Wahler, R. G., Winkle, G. H., Peterson, R. E., & Morrison, D. C. "Mothers as behavior therapists for their own children." *Behaviour Research and Therapy*, 1965, 3, 113-124.

Ward, M. H. & Baker, B. L. "Reinforcement therapy in the classroom." *Journal of Applied Behavior Analysis*, 1968, 1, 323-328.

Wetzel, R. J., Baker, J., Roney, M., & Martin, M. "Out-patient treatment of autistic behavior." *Behaviour Research and Therapy*, 1966, 4, 169-177.

Whaley, D. L. & Malott, R. W. *Elementary Principles of Behavior.* New York: Appleton-Century-Crofts, 1971.

Williams, C. D. "The elimination of tantrum behavior by extinction procedures." *Journal of Abnormal and Social Psychology*, 1959, 59, 269.

Wolf, M., Risley, T., & Mees, H. "Application of operant conditioning procedures to the behavior problems of an autistic child." *Behaviour Research and Therapy*, 1964, 1, 305-312.

Zeilberger, J., Sampsen, E., & Sloane, H. N., Jr. "Modification of a child's problem behaviors in the home with the mother as therapist." *Journal of Applied Behavior Analysis*, 1968, 1, 47-53.

Zimmerman, E. H., Zimmerman, J., & Russell, C. D. "Differential effects of token reinforcement on instruction-following behavior in retarded students instructed as a group." *Journal of Applied Behavior Analysis*, 1969, 2, 101-112.

Many people have called for a reconceptualization of mental illness and mental health in the broader context of everyday life. In the paper that follows, Atthowe focuses on the delivery of mental health services which are all-encompassing and not fragmented, and provides a framework for this reconceptualization. It is his thesis that "treatment", by itself, does not go nearly far enough in actually solving the problems that the patient brings to the treatment setting. This failure is frequently associated with the therapist's complete neglect of critical components of the social system and his total ignorance of how to develop preventive programs and environments. Consequently, Atthowe believes that we must start with clearly defined statements about "mental illness". He reviews several sets of problems associated with our initial assumptions about our treatment mission.

Atthowe believes we must provide comprehensive community human services because it is the only way we can effectively deal with realities of people's problems in living. However, comprehensive community services require a new breed of professional and a new delivery system for services. Both of these changes create complicated problems for the "mental health establishment" and will force professionals to seriously re-evaluate their contributions. In addition, Atthowe calls for a new technology of behavioral intervention which includes an extension of social learning principles and involves developing an adequate data base for both planning and operational decision making.

John M. Atthowe, Jr.

CHAPTER 9

Behavioral Innovation: An All-Encompassing System of Intervention

Mental health and mental health services have a long past but a relatively short history. We would be unduly self-centered if we equated mental health care with the emergence of the mental health professional. The domain of mental health and mental illness has always been a major part of the domain of living. Men and women from all walks of life and from all points of view have concerned themselves in one way or another with the problems discussed in this conference book. But, it has largely been the mental health professional who has defined how mental health and illness is to be viewed and how mental health services are to be delivered. However, our definitions, our viewpoints and the services we have delivered have varied widely.

The domain of mental health and mental illness has stressed from time to time a somatotherapeutic orientation, in which physical and chemical etiologies and treatment dominate; a psychotherapeutic orientation, in which psychological (inner dynamics) and behavioral (outer reactions) models of etiology and treatment are emphasized; and more recently, a sociotherapeutic orientation, in which social and community factors (the ecology in which the person lives) are stressed (e.g., Ehrlich & Sabshin, 1964). Yet, all of these approaches share one important feature in common. They have all delimited the domain of mental health to psychotherapy (individual and group) at the therapist's place of business. However, in recent years prevention and aftercare (maintenance) are becoming areas of concern largely due to the ineffectiveness and inefficiency in the delivery of mental health services (e.g., Atthowe, 1973a; Cowen, 1973).

The major theses of this paper are that mental illness and mental health must be viewed in the broader context of everyday living if we are to be successful in dealing with prevention and maintenance, and that the delivery of mental health services must be all-encompassing. By all-encompassing I mean not only that treatment (i.e., individual and group or family therapy, chemotherapy, somatotherapy, assessment, retraining, crisis intervention, etc.) must be undertaken, but that maintenance of the effects of treatment and of mental health in general (preventive maintenance) must also be undertaken.

Furthermore, in order to encompass the entire domain of mental illness and mental health, services must involve the development of preventive programs and environments, and steps must be designed to insure the perpetuation of these prevention systems. Otherwise, the effects of treatment, no matter how successful, will continue to be short-lived (Atthowe, 1973a) and the problems of living and human distress will remain largely untouched.

It is time that we stop flogging the medical, the behavioral, the psychoanalytic or the humanistic model of mental illness. All have their place, albeit very limited, in an all-encompassing model of mental illness *and* mental health. To arrive at a more comprehensive and more effective approach to mental illness and mental health, we must look first at how society, the customers, view both mental illness and mental health, and the mental-health delivery system.

It is what the customers, not the professionals, of our service do or fail to do that defines the problems that exist. Only then can we develop effective delivery systems which would involve treatment, maintenance, prevention and perpetuation. However, many obstacles stand in our way. Some of these obstacles reside within the mental health guild (London, 1964) itself, some within our current system of delivery of service, some within the community (the maintenance and prevention milieu), and some within the system of beliefs and values that define our culture.

In this paper I will focus first on the concepts of mental illness and mental health and the effects of current forms of treatment. Next I will deal with a more expanded concept of mental health, especially positive mental health or human service and some of the goals or values upon which a human service program might be based. Thirdly, I will attempt to define a more comprehensive model of intervention and some procedures for developing, utilizing and evaluating this model.

THE CONCEPT OF MENTAL ILLNESS AND MENTAL HEALTH

The concepts of mental illness and mental health are often used interchangeably; logically, however, health is the complement of illness. Therefore, a program of help or care must include both concepts. Newer concepts of service have increasingly emphasized the idea of health (e.g., preventive programs of mental health, the promotion of mental health, the maintenance of mental health, the perpetuation of mental health programs, etc.). Treatment, on the other hand, assumes a deficit — a lack of mental health, or mental illness. In addition, mental illness is viewed as a negative term and as such may help to turn people away from mental-health professionals and services. For that reason, it is imperative to look at how Americans view mental health and mental illness.

How Americans View Mental Illness

There seems to be little known about how Americans view positive mental health. Mental health and mental illness seem to be synonymous in the eyes of the general public. In general, mental illness is seen by the American public as a serious and fairly frequent occurrence, but at the same time a very negative one.

In one of the most representative nationwide surveys of psychological problems, it was found that one out of every seven persons, 21 years of age or older, admitted to having sought psychological help for his or her problems; and, in addition, one in every four adults admitted that he or she needed psychological help (Gurin, Veroff and Feld, 1960). However, less than twenty percent of those who sought help actually visited mental health professionals. In other words, fifteen years ago less than three percent of the nation's adult population sought out the services of the mental health profession. But what of the other eighty percent who sought help elsewhere? Where did they go?

The real purveyors of mental health turn out to be clergymen (42 percent) and physicians (29 percent), and in rural and "high-risk" areas, the public health nurse or community worker. The message rings loud and clear; people with psychological problems, people in distress, more often avoid those professionals whom society has labeled as the deliverers of mental health services. We must ask ourselves, why?

In part the avoidance of mental health professionals by persons with psychological problems is due to their reinforced realization that rejection, by society, of those who seek out mental health professionals is much greater than of those who consult clergymen and medical practitioners (e.g., Phillips, 1964, white married women; and Yamamoto and Dizney, 1967, student teachers). Studies regarding people's opinions about mental illness and mental patients, and the opinions of mental patients themselves, are predominantly negative and rejecting of mental illness. In a recent review of opinions about mental illness, Rabkin (1972) summarizes these results as follows:

> In summary, mental patients are as negative in their opinions about mental illness and the "insane" as the general public. They appreciate the value of psychiatric hospitals less than hospital staff.

In short, society's attitudes and actions toward the mentally or psychologically ill are those of avoidance and rejection. Informing the general public about the nature of mental illness does not seem to have altered these attitudes or behavior (Rabkin, 1972). In fact, promoting positive attitudes toward mental illness may even boomerang (Cumming and Cumming, 1957).

It seems that changing attitudes in isolation, and not taking into account the system of beliefs and attitudes of which they are a part, can lead to dire results. People's opinions and beliefs regarding mental illness and mental health are embedded into their overall belief system which, in turn, is dependent upon the system of opinions and beliefs held by one's peer reference.

Mental health professionals, as a special reference group, seem to see more mental illness than people will admit to. For example, in both the Midtown Manhattan (Srole, et al., 1962) and the Stirling County (Leighton, et al., 1963) studies, psychiatrists rated 60 percent of the population studied, moderately or seriously bothered by symptoms of mental disorders. It also has been estimated, based largely on school reports that approximately one in every four-or-less elementary school children has an adjustment problem (Glidewell and Swallow, 1968), and that the incidence is much greater among the inner-city poor.

Some of the difference in evaluation between professionals and the general population may be due to a much narrower view of what constitutes serious mental illness on the part of the average man-in-the-street. This is especially true for those who are older and in the lower socioeconomic classes (Dohrenwend and Chin-Shong, 1967). It seems likely that mental health professionals are only associated with very serious disturbances.

Whatever the answer to this question, mental health professionals must ask themselves if their behavior is contributing to society's rather limited and negative attitude toward mental illness.

Another factor which must be considered is the well-documented correlation between the type of treatment and the social class of the patient (Lorion, 1973). Not only do the lower socio-economic classes have the highest incidence of psychiatric problems, but they also are the least likely to receive professional help (Srole, et al., 1962), and the help given them has been the least successful (Cowen, 1973). Lower socio-economic status patients ". . . are significantly less likely to be assigned to individual treatment, and if assigned, are more likely to terminate prematurely" (Lorion, 1973).

Whatever the reason, a substantial portion of the population (the lower class) is having a hard time acquiring mental health service from those giving it (the middle class) when they seek such service. It is clear that we cannot afford to view the person seeking help as a separate entity. We must consider his background and the environment from which he came and to which he will return. Furthermore, the treatment team (the middle class professionals) is as much a part of the treatment as the procedures themselves, and in some cases even more. If treatment is to be maximally effective, we must take steps to maximize the placebo effects, the demand characteristics of the intervention and the expectations of success for all segments of the population. Attitudes toward mental illness, toward positive mental health and the delivery of mental health services, must be realistically changed.

The Problems of Treatment

If all of the individuals who supposedly need or desire help were to seek it, the demand would far exceed presently available resources, if those resources continued to be delivered as they are at present. However, serious emotional problems are only the tip of the mental-health iceberg. One of the mandates of the Community Mental Health Movement was to develop centers which would also handle problems in living. A new breed of mental health professional would be trained to actively intervene in problems resulting from poor housing, unemployment, a lack of money, food, and the basic necessities of life, and social and political oppression. If such centers actually came into existence, the problems associated with the delivery of mental health services would be overwhelming indeed. However, as Nader and his workers (Holden, 1972) attest, the approximately 325 mental health centers in operation "offer mostly a collection of traditional clinical services" which largely remain "inaccessible or irrelevant to large segments of the community".

Even if we do not subscribe to this broad mandate for the operation of Community Mental Health Centers, the problem of "too little" and less than effective help remains. The outcome of the treatment of non-institutionalized mental illness, though it may be "modestly positive" (Bergin, 1971) in the short-run, is hardly demonstrable in the long-run. Furthermore, the outcome of treatment of serious mental or behavioral disorders (i.e., institutionalized individuals) is generally much worse in both short-term and long-term evaluations. A recent review of current psychiatric rehabilitation practices resulted in the following conclusions (Anthony, et al., 1972):

> 1. Traditional methods of treating hospitalized psychiatric patients, including individual therapy, group therapy, work therapy, and drug therapy, do not differentially affect the discharged patients' community functioning as measured by recidivism and post-hospital employment.
> 2. Aftercare clinics and other forms of moderate community support reduce recidivism.
> 3. Various types of transitional facilities are successful in reducing recidivism but have demonstrated little effect on enabling the patient to function independently in the community as measured by post-hospital employment, [numbering ours].

Chronic and serious disturbances generally have not responded to inpatient treatment. Even outpatient services involving less serious problems have failed to demonstrate persistent mental health. Long-term follow-up or longitudinal studies, though few and hard to evaluate, indicate that individual or group psychotherapy is not enough. Treatment plans must be expanded to include maintenance systems designed not only to reduce recidivism by supportive and transitional living arrangements, but also to maximize employment, education, and community-related activities through transitional training and programmed community activity (e.g., Atthowe, 1973a; Fairweather, et al., 1969).

Preventive maintenance in the "real world" is a necessity if we are to maximize health benefits and reduce the cost of treating illness. The cost to society of keeping one person in a State Mental Institution in New Jersey is approximately $14,200 per year. The cost of treatment is approximately $9,500, and the loss of productivity by keeping someone out of the labor market is $4,700. However, discharging 74 percent of a cohort of 92 patients resulted in a 63 percent reduction in the overall figure over a two-year period. Of the 37 percent who stayed out of the hospital, the percent savings to society per year over a two-year period was approximately 90 percent or $12,780 (Johnson and Pollack, 1973).

For the minor problems of living, preventive maintenance may more effectively take the form of establishing transitional clubs, and recreational and social centers, and working directly or indirectly with family units, educational systems, industrial organizations and "high-risk" or inner-city environments. In short, treatment is not enough; other systems of delivery must become involved — systems of maintenance and prevention that are built into the social system of the community.

BEHAVIOR MODIFICATION, SYSTEMS THEORY, AND COMMUNITY MENTAL HEALTH

In the short lifetime of behavior therapy and behavior modification, extensive changes have occurred. Largely derived from conditioning theory in the work of Eysenck (1959), Wolpe (1958), and Skinner (1953), behavior therapy, and behavior modification in particular, have expanded into the realm of social learning (Bandura, 1969; Patterson, 1969), and behavior influence (Krasner and Ullmann, 1973; Ullmann and Krasner, 1969). Behavior modification in its community orientation has evolved as a special case of the influence process.

As community mental health programs have grown, new roles have emerged for the behavior modifier. Treatment has broadened and has gone into the community or natural environment (e.g., Tharp and Wetzel, 1969). As Patterson (1971) points out, the dispensers of reinforcement for the deviant individual are those people who constitute the individual's social environment.

> Presumably, it is they who shape and maintain his deviant behavior . . . then it would follow that the focus for intervention should be that of attempting to modify the dispensers who provide the contingencies. (Patterson, 1971)

In other words, the person's deviant or ineffectual behavior is more often triggered, and maintained by the social system of which he is a part. However, the deviant person is usually a member of more than one social system; furthermore, when therapeutic intervention (treatment) takes place it normally involves a number of different professional guilds and care systems.

Each social system has its own contingencies, its own special instigating conditions and its own unique reinforcers. In some cases these contingent relations are quite explicit, as rules or codes of conduct, but more often they are implicit, as unwritten and informal status hierarchies, etc. In any case, the mental health worker must take into account the relevant social systems surrounding the patient both inside and outside of the treatment center if we are to effectively modify human behavior. And, in many instances, the social system(s) rather than the individual should be the target of our intervention.

The effects of the different, and often competing, mental health delivery systems on treatment and the case for a more comprehensive system of intervention can be demonstrated by a personal case history. In 1962, together with Len Krasner, I started a behavior modification unit at the V.A. Hospital in Palo Alto. Due to the controversial nature and the newness of the project, we were assigned an 86-bed chronic back ward (a ward no one wanted) in which the average length of hospitalization was 22 years. Various techniques, such as relaxation training, systematic desensitization, aversive conditioning, and reinforcement therapy were tried with little or no success. As positive reinforcement seemed to be the most effective procedure, we decided to develop a token economy for the entire ward.

The development of a ward or unit token program was begun. At first we had to train the already-existing staff. This was not easy and necessitated a one-year (as opposed to a one- or two-month) "operant" baseline period while staff attitudes were changed. During this period there were frequent meetings with the ward staff in an attempt to establish a program in which everyone would work together. At this stage it became apparent that the success of a token, or any complete hospital program was dependent upon the cooperation of the entire ward staff, in particular the nursing service (nurses and aides). Each professional service had its own implicit, as well as explicit, social system. The very fact that the head of the psychology service, Dr. Kennelly, was aware of these contingencies, and intervened with the heads of the various services, enabled us to develop a viable program.

After some staff changes and the selection of a few key individuals, the program started. It soon became apparent that in order for the program to be really effective other off-the-ward service groups had to be included in the program. What was done on the ward could easily be undone by occupational and recreational therapists, by maintenance workers, and by other hospital employees and patients. If we were to develop and maintain an effective token program, we had to include the entire hospital milieu within our treatment plans. This meant that we had to educate and influence others, especially those within the social systems within which our patients interacted. It became clear that the persons who controlled the contingencies and dispensed the reinforcements for our patient population were many and varied. Therefore, the primary target of our program turned out to be those who dispensed the reinforcements and controlled the contingencies (the hospital personnel). Whether we realized it or not, we were utilizing systems theory as well as behavior modification procedures. Our goals were to shape the relevant hospital social systems as well as the patients themselves.

Staying on the ward on a 24-hour basis provided me with a much better picture of the real operating reinforcers and contingencies, especially those that developed after five p.m. And, as the program developed and our evaluative procedures blossomed, we found that certain environmental events had tremendous impact on patient behavior. For example, the Christmas season markedly disrupted the performance of most patients in a negative direction.

We found that operating a token program was more than merely following a set of rules and procedures. The effective administration of backup reinforcers necessitated knowing the social systems operating in general or at the time of the administration, be these intra- or extra-hospital or intra-individual. If some individuals earned either too many or too few tokens relative to their utilization or expenditures, their performance generally was negatively affected.

A treatment program invariably involves the "art" of shaping (see Sidman, 1962). When, with the advent of a new program administrator, weekly staff meetings were greatly reduced and informal "coffee conferences" eliminated, a decided drop in performance occurred along with a drop in staff morale. Thus, we discovered, a program involving mediators of change must also apply shaping principles to the mediators. All of us need to be reinforced from time to time for what we do.

Another important factor eventually became apparent. When a patient would leave the hospital, it was not long before he returned. Like all hospital treatment programs, recidivism was too high. Taking a cue from a concurrently operating program (Fairweather, 1969), we realized we had to prepare and deal with the post-hospital environment if we were to fulfill our charge of maximizing mental health services. Therefore, we belatedly took steps to shape individuals into leaving the hospital (a major transition in their lives), and residing and maintaining themselves in the world outside (see Atthowe, 1973a). Comprehensive care dictated the inclusion of an additional dimension to the treatment milieu, the maintaining milieu with its subsequent social systems.

The above excursion into the awakening of my concerns with the social-systems operative in the delivery of comprehensive and more effective mental health care can be generalized to mental health care in general. The rest of the paper will be directed toward this end.

POSITIVE MENTAL HEALTH (HUMAN SERVICES)

Mental illness and its complement, mental health, more often than not generate avoidance behavior (rejection) of available mental health services and mental patients. Consequently, if we are about to revolutionize (see Cowen, 1973) the mental health field by moving into the social and community sphere, shouldn't we consider what services the public, themselves, want, and the possibility of changing the name of the services we deliver? As the Nader report (Holden, 1972) suggests, we might better call our centers human service centers, and our practitioners human service workers.

However, we are stepping on the toes of established disciplines when we do this. The guilds (see London, 1964) would hardly accept such a state of affairs. However, a human service worker might not have to fight the medical model, accreditation agencies, nor the insurance companies for the right to perform human services; separate disciplines with special territorial rights might disappear; and individuals seeking human services might be more able to get comprehensive services plus insurance coverage for basic living problems. There is no doubt that some problems call for chemotherapy or medical or neurological skill, and others, individual and group psychotherapy. Even in these cases, however, social support and maintenance are necessary. [All types of services (medical, psychotherapy and social-community intervention) are necessary for long-term effectiveness.] Internists, neurologists, pediatricians, etc., also would be more effective if they were associated with all-encompassing human service programs. We have for too long concerned ourselves with individual expertise in the health fields. Cooperative and joint expertise is becoming a necessity if we are to resolve the persistent problems we face.

Comprehensive Human Services

Human problems, needless to say, are many and varied and generally require comprehensive attention; including, medical, psychological, social, and ecological. In some instances, one or the other might be emphasized, but comprehensive treatment generally would include all of these services. For example, a chronic physical illness is not merely a physical illness, per se; it is, or it becomes, a dependency on others. Most patients in extended-care facilities have stabilized illnesses and require care for general dependency rather than for their illness (Salmon, et al., 1966).

Different social settings provide different occasions for experiencing different types of crises (frustrations and conflict), and learning different ways of behaving to them. Instigating conditions from different sources, reinforcing conditions of different magnitudes, contingencies of an almost unlimited nature, and idiosyncratic reactions to each provide the broad, ecological basis for human behavior and human problems. If we fail to include social and community systems within our model, our problems will reappear in most instances.

The case for more social and community intervention, and against traditional treatment, is well documented by Cowen (1973) in his review of community mental health programs. In his concluding remarks, Cowen (1973) has this to say:

> When widely accepted frameworks do not effectively handle problems within their mandated scope, and as new relevant problems are identified that they fail to address, it is time to re-examine assumptions and to consider alternatives. . . . Cold evaluation of past dominant Mental Health approaches invites the criticisms that . . . they have not provided the manpower and resources needed to cope with evident problems and latent needs, . . . their pivotal techniques have had, at best, limited effectiveness; and they either do not reach, or are inimical to, major segments of the population urgently requiring help.

As for social and community interventions, Cowen (1973) adds:

> . . . the evolving Social-Community framework offers a genuine alternative to prior dominant Mental Health approaches. It is active rather than passive and accords far greater importance to prevention than to repair. Its key components include analysis and modification of social systems, including engineering environments, . . . that maximize adaptation. [It stresses] . . . early childhood intervention, crisis intervention,

> and consultation. . . . As such, it focuses attention on the person-shaping attributes of communities and their primary social institutions.

If we agree that human services (i.e., the delivery of mental health services) should be comprehensive, we must then ask the question: Is it possible? Most people will say that comprehensive intervention is unfeasible, inefficient, and probably impossible to accomplish. This would be true if we continued to equate intervention with treatment. It is the contention of this paper that the delivery of mental health services is going on all around us all the time. However, we must learn how to harness these resources. The occasions for human problems and misery are even more prevalent. We cannot stand idly by; we must actively intervene. We must develop and perpetuate both maintaining and prevention milieu with the natural environment. The creditability gap between what we are now doing and the results, and between those who could benefit from these services and are not receiving them, is widening, and will widen even more. Third party insurance claims, legal accountability in the courts, peer evaluation, and competition for the few remaining dollars will not allow these gaps to go unnoticed.

Comprehensive human services or positive mental health requires a new breed of professional and professional training. We need the generalist as much as we need the specialist. How this future human service worker will be trained will play a major role in the development of this approach. First of all, this worker must be sensitive to, and discriminating of, the problems of living. Obviously he or she cannot be aware of all the problems faced by all human beings. Consequently the human service worker should be trained in the ways of different people and their environments, in cultural anthropology, environmental sociology, individual differences and in behavioral genetics. He or she should be taught to think in terms of gains rather than cures, and the interdependence of people upon each other and upon their environment (e.g., general systems approach). He or she should be aware of the effectiveness, or lack of it, in the current systems of treatment (e.g., somatotherapy, psychotherapy and sociotherapy). He or she should be taught the relevance of preventive maintenance for long-term effectiveness, and the necessity for perpetuating these environmental interventions. He or she should be concerned with developing acceptance and participation by the community in the maintenance, and thus the promotion, of positive mental health.

The human service worker should be taught the techniques of organizing prevention or promotion programs within groups of primary care agents (e.g., clergymen, teachers, lawyers, physicians, public health workers, etc.) and service agencies. Additional preventive or promotional measures such as the training of others in crisis intervention and human service philosophy and technology would be part of the human service worker's education. The human service worker should also be aware of the technology of attitude change and behavior modification and influence so that his promotional, preventive and maintenance campaigns would be successful. And, he or she should be aware of the vital role of key political figures and economic institutions in controlling the reinforcements and in establishing the contingencies of daily living. In this latter sense, the worker should receive training in law and civil procedures, economics, and political theory and action.

Instead of intensive training in the theories of personality development and psychopathology which in turn limits our perspective, a comprehensive perspective would provide extensive training and familiarization with the various social systems that critically affect human behavior. Aside from those who wish to make their lives the exploration of human personality, human service workers would need only know what the major theoretical trends are and what works. Then, as human service centers grow they would take on more and more the function of in-service training and research centers.

Human Services, Human Values and Attitude Change

Positive mental health assumes one thing is better than another, that some course of action, some goals, are more valued than others. Such values, implicit and explicit, have appeared throughout these pages. One assumption, however, that underlies social and community intervention is that a healthier environment sets the occasion for healthier individuals, and healthier individuals in turn will reduce the myriad of frustrating and conflicting occasions within their environments. Such a view does not minimize the biological basis of human suffering; it adds additional substance. A further assumption is that human behavior does not occur in a vacuum; it occurs within the context of many on-going intrapersonal, interpersonal, social and cultural systems of behavior. If we fail to take into account each of these functioning sub-systems, our analysis and, thus our intervention, would be less than adequate.

An active approach to mental health assumes a new set of attitudes, a new paradigm. A medical model or its behavioral therapy, psychoanalytic or humanistic offshoots, is a passive model in which those desiring service seek out the professional. On the other hand, the public health model stresses pathology and does not emphasize the interrelationship between the different on-going systems of influence of control, and human problems.

One of the major innovations in the treatment of the seriously ill has been an attitudinal one: the elimination of the concept of "cure", and the substitution of the concept "gain" or change. In a like manner, the concept of treatment is better subsumed under the all-encompassing rubric of comprehensive intervention in which intervention becomes an active search for ways to bring about positive mental health and the maintenance of ex-patients within the natural environment, in addition to the traditional forms of treatment. A treatment plan should always include the person's natural environment as part of the professional's thinking; consequently, mental health or human service centers must also think in terms of preventive projects and promotional programs.

Positive mental health or effective human services would be predicated on the notion of basic or common reinforcers. The issue becomes one of defining the most basic reinforcers shared in common, and those contingencies that control or influence that class of behavior we call positive mental health. However, positive mental health is intricately woven into the activities of daily living. Therefore, we must seek out those reinforcers and contingencies that influence or control all of human behavior.

Skinner (1971) in his blueprint for a better world points out that human behavior is always controlled. The oppressed individual would readily concur. Accordingly, Skinner would see one of man's most basic goals as the avoidance of aversive controls. The American judicial system is predicated on just such a notion. But the legal system, alone, cannot undo the extensive and intensive impact of aversive control. However, as the recent consumer and the older union movements have demonstrated, mass effort can successfully influence aversive consequences. For example, one shopper cannot bring the prices of a food commodity down, but a group of shoppers can, if practices were to persist. The larger the group of shoppers the greater the impact.

The point is that if we wish to influence society's power over economic structure of a community we need to organize large segments of that community and maintain that organization. Human beings by banding together and sharing certain goals and avenues of attainment, can minimize the frustrations and the conflicts so evident in society. Obviously, frustrations and conflicts cannot be wholly eliminated, but we can make inroads by promoting positive mental health, by developing preventive programs, by advocating changes in the social system of which our client is a part, and by helping to create a healthier system of values. The individual in distress is one of our own. The community should not cast out everyone who deviates. We are all on the same spaceship headed for, we hope, a better quality of life.

Human beings have another option open to them. That is, man is capable of reacting to his own behavior and its consequences. Human beings still have the option of reacting to aversive controls by not reacting, by "turning the other cheek", "tuning out", taking it all in stride. Not reacting or only damply reacting to aversive consequences or potential aversive consequences are examples of coping or adaptive behaviors. It would be foolish not to advocate these latter coping behaviors. However, reactions shared in common by a large number of individuals or by a few "key" persons can influence the way in which reinforcers are attained, or aversive consequences avoided, and ultimately, influence what is valued (i.e., what man's basic reinforcers should be).

The notion inherent in this paper is that a new set of actions and a new class of values regarding mental health is emerging. We have described this action; its goals, however, remain somewhat vague. When we speak of prevention, maintenance and perpetuation, we are tampering with on-going social systems; in essence, we are re-directing some of the goals of these social systems. Should these goals be more than the avoidance of serious and persistent problems and psychopathology?

Some personality theorists ascribe to an all-inclusive social goal as the ideal goal of society, such as love or concern for the human race (Allport, 1961; Fromm, 1956; Maslow, 1954). Others would ascribe to a more idiosyncratic goal, such as happiness (e.g., Skinner, 1971; Wilson, 1967). Rawls (1971) in his *Theory of Justice* proposes that every person should have an equal opportunity to achieve happiness as long as this opportunity does not conflict with the happiness of anyone else. Inequality is only justified by Rawls (1971) if it is necessary for a shared common purpose, such as to benefit the least-advantaged or least-effective person.

It seems to me that we must redefine the concepts of respect and self-esteem, such as the sharing of a common goal, the goal being respect and concern for others rather than power or wealth; then the more respect a person showed for others, the greater would be his respect within the community as well as his self-esteem. In the human service area greater respect for others could be supported and reinforced within that community's social system as part of the prevention design of the service center. Obviously, we are dealing with abstract ideas; however, there is no reason why the values of a society, which are man-made, cannot be operationalized and dealt with. A few disillusioned groups of young people are beginning to do this in individual communes. I am proposing that this is a legitimate function of human service centers and that a healthier society is a necessary part of comprehensive service.

A COMPREHENSIVE MODEL OF INTERVENTION

This paper has attempted to demonstrate the necessity as well as the urgency for the development of a comprehensive model for the delivery of mental health services. If we are to be a socially responsible group of professionals, and if we take seriously the mandate for the accountability of our actions, then we must move into the realm of prevention and maintenance. This model of prevention, treatment, maintenance and the perpetuation of these effects, we will call *behavioral innovation* (Atthowe, 1973a). It is behavioral in that it stresses what people do or fail to do. Every model that is carried out is at least a behavioral model. The model is concerned with modifying or making a change in the established order of things. In this sense it is innovative. In promoting positive mental health we are called upon to influence and to change the system of beliefs, values and activities of a client, a group of clients, an organization or institution, a community and the culture of which the community is a part; in this sense the model is a general systems model.

As we have mentioned previously (Atthowe, 1973a), planning behavioral innovation is a tremendous responsibility. It always has and always will be with us. However, up to now social change and behavioral innovation have been carried out by a few visible, select politicians and a larger number of non-visible industrial or business leaders and power wielders. Innovation within the natural environment, no more than innovation within the typical mental health center, requires the sharing of a common goal. The overt and covert agents of control or influence are largely concerned with maintaining the status quo in whatever setting they may be found; and, in this regard, these agents of control have long been aware of the very low correlation between verbal expression and overt behavior. As long as social planning is merely talk, no one cares; thus the doctrine of behavioral innovation can be considered explosive.

Comprehensive planning and service is beyond the scope of most of us. Behavioral innovation requires a service team consisting of specialists from many different professions (the learned professions: social, behavioral, biological, and environmental sciences, law, urban planners, education and business management), and walks of life (the experienced professions: the inner-city minorities, the poor, the suburbanite, the blue-collar worker, the white-collar worker, etc.). However, the key members of the team would be the human service workers described previously.

The human service workers would be the main mediators or coordinators of service. The notion of a non-professional worker is a misnomer. If the coordinator of service is untrained and unselected, service would no doubt suffer. The most effective coordinators and mediators of training would be selected individuals who are found somewhere upon a career ladder from the early trainee through the worker, the technician, the specialist to the generalist. One such series is the Arizona State Hospital Mental Health Series which starts at Grade 3 (Mental Health Worker Trainee) and progresses through Grade 15 (Mental Health Specialist II, Nursing). The minimum requirements for Grade 15 would be a bachelor's degree and two years experience in the next lower series. I believe the next higher grade should re-introduce into the mental health field, with training equivalent to that of a master's degree, the generalist.

The Technology of Behavioral Innovation

Behavioral innovation is an extension of social learning (e.g., Patterson, 1971) and operant learning (e.g., Tharp and Wetzel, 1969) principles to the natural environment, plus the influence of the notions of behavior influence as expounded in Krasner and Ullmann (1973).

Two other approaches contribute to the conceptual basis of behavioral innovation. One is the creation of a community of treatment as exemplified in the work of Fairweather and his associates (1964; 1969). The second is the systems approach to community mental health as seen in the writings of Cowen (1973). Finally, behavioral innovation represents a phenomena of growth encompassing some of the ideas derived from token economies (e.g., Atthowe, 1973b; Krasner and Atthowe, 1971), planned environments (e.g., Atthowe, 1973a; Fairweather, et al., 1969; Krasner and Krasner, 1973; McDonough, 1973), and social philosophers such as G. H. Mead (1934) and Skinner (1971).

Let us assume, we are charged with opening a human service center. As in any analysis, the *first* step involves gathering an adequate data base from which to proceed. This phase requires observation of just what is happening, the contingencies, the reinforcers, and the critical agents of influence. Epidemiology would be another way of establishing mental health and human service base rates and stable baselines for that community.

The *second* step, an analogy to Weed's (1971) system, would be the tentative description of the problems to be further defined and validated. This would involve additional community analysis and hypothesis testing.

The *third* step would be the development of procedures by which the center would be accepted within the total community.

The *fourth* step would involve the development of preventive and preventive-maintenance systems by working with already-established care and helping groups, such as the primary purveyors of care (clergy, teachers, parents, physicians, lawyers), the secondary purveyors (social agencies, courts, etc.) and the main controlling agents (political, industrial and indigenous leaders).

The *fifth* step would be the opening of the human service center which would include the various components of help and care: walk-in crisis intervention centers within each of the functional communities; outreach teams emanating from these community centers who play the role of patient or client, advocate and counselor; a backup emergency service and inpatient services; a partial hospital complex including day, evening and night programs; a rehabilitation program encompassing a 24-hour operating hospital, transitional work (e.g., gas stations, typists, etc.) placement areas, community clubs, and transitional living arrangements; specialized outpatient services which would be offered to all age groups; remedial school programs and day hospitalization for school children; special programs to deal with drug and alcohol problems; a prevention and promotion unit concerned both with coordinating and establishing prevention programs in the community using contingency contracting, and educational and promotional projects stressing acts of commitments rather than lectures, per se. Another direct service function would be performed by the outcome evaluation, accountability and automated feedback section, in our center, REAR (Research, Evaluation, Accountability, and Records). This latter unit would help to determine if our undertaking is really worthwhile.

The *sixth* step would be the development of the comprehensive intervention plan. The problems would be defined and the goals of intervention stated in the form of terminal behaviors or activities (Houts & Scott, 1972) for both long-term and short-term objectives. The actual intervention procedures would be spelled out in detail. Treatment and preventive-maintenance, as well as the development of procedures designed to perpetuate the prevention program, would be part of every intervention plan.

The *seventh* step would be the carrying out of the intervention plans. Progress notes (the degree of attainment of the short-term and possibly long-term goals, changes in the problems, and goals or plans), and immediate feedback via automated record keeping and the utilization review process, would be continually updated. A continuous flow or operations chart could be developed not only for people seeking help (the patient's medical record) but for community prevention projects (the problem-oriented operational chart), for service units, for individual service workers and for the center as a whole.

The *eighth* step involves the termination of the intervention plans and the taking of steps to insure that the progress is maintained and enhanced and that the prevention or maintenance program is perpetuated in the community. Like any program designed to produce enduring changes, the fading-out of the service center's influence should be gradual as the natural reinforcers and controlling agents in the community take over. The ultimate goal of any positive service program is the development of self-sufficiency. The person or the preventive program must come to stand on its own two feet. Training programs within the community, and as part of the development of the prevention projects themselves, will help to insure the perpetuation of the programs. It should also be noted that persons could enter and leave this comprehensive system at various points.

Evaluation and Accountability

We have reached a point in time where a gap in credibility exists in all fields and endeavors. This is becoming more and more apparent in mental health. There is no doubt that mental health is in need of accountability. Outcome studies, as we have pointed out, have not been good. Another force that is creating a need for accountability in our planning procedures of treatment and goal-setting is the law, or legal accountability (Wexler, 1973). The Alabama decision (see Wyatt vs.

Stickney in Wexler, 1973), and subsequent decisions indicate that we better take our social responsibility and record-keeping seriously and justify what we do or do not do (i.e., become accountable).

Computerized record-keeping (e.g., Glueck and Stroebel, 1969) and the adaptation of problem-oriented medical records to the mental health arena (e.g., Grant and Maletzky, 1972) have made the clinical record a service tool in which immediate feedback and self-correction is built into the service operations, and in which goal attainment can be continually evaluated. Such records also provide ready training and well-supervised service experience. If we are to train a myriad of new workers in the 70's, we need to utilize every avenue of training available.

Finally, we must zero-in on the problem of outcome or program effectiveness. As pointed out earlier, there is a great monetary savings in reducing inpatient care, especially in dealing with the chronic patient. Not only is there a marked cost benefit from reducing recidivism but even more so from creating post-hospital employment. *Cost-effectiveness* analysis is becoming one of the more important alternatives in evaluating program or treatment effectiveness, especially when recovery is not complete.

In the Johnson and Pollack study (1973), reported earlier, the total cost to society of institutional care was broken down as follows: 1) the cost of treatment for 92 inpatients was approximately $9500 per patient per year; 2) loss of income through decreased productivity was approximately $4700, totaling $14,200 per patient each year. Over a two-year period, 25 patients remained in the hospital, 34 were discharged and remained out of the hospital, and 33 were repeatedly in and out of the hospital. The actual cost to society (i.e., the cost of dependency) was reduced to $5300 per patient each year. The cost of treatment dropped from $9500 to $3400, and the loss due to a lack of productivity dropped from $4700 to $1800. Outpatient treatment when added for some patients was minimal ($45 per year for each patient). Costs of shelter and food were not included since the costs would have been incurred whether or not the person had been hospitalized.

Such an analysis points out the large reduction in dependency which can be produced by post-hospital employment. As is often the case, it is the marginal man with minimal skills who is institutionalized. Training or habilitation programs would not only increase the individual's self-sufficiency and self-respect but would also markedly reduce his dependency and society's costs. A cost-benefit analysis provides administrators and budget controllers a basis for discriminating between programs of differing effectiveness and efficiency. However, one important factor is typically missing from this equation. This is the psychological cost to the person or persons. How do we account for intra-individual costs?

Intra-individual costs have been analyzed by *patient or therapist ratings* with minor success. Such studies can be criticized for the unreliability of the results and the low correlations between verbal statements of improvement and actual behavior (Bergin, 1971; Eysenck, 1966). Client satisfaction, therapist satisfaction and "significant others" satisfaction (e.g., parents, relatives, school personnel, etc.) are important though unreliable measures. However, the more tangible the conditions rated, the greater the reliability.

Some studies indicate a considerably higher validity and reliability from ratings by relatives than those of therapists. The question ultimately becomes how best to explicate a person's subjective feelings. Symptom checklists are subject to the same criticisms as self or therapist ratings, largely due to the vagueness and unreliability of the items, and the client's or therapist's biases. Again the more behavioral the items and the more relevant they are to the problems considered, the greater their validity and reliability. As pointed out earlier, the general public and mental health professionals frequently disagree as to what constitutes a real problem.

Many of these issues can be minimized by utilizing the *problem-oriented medical record* (Weed, 1971) for patients and problem-oriented operational procedures for program development and evaluation. The essence of the problem-oriented record is a description in actual behavior (overt, verbal or physiological) of the problem as seen by the patient and/or the service team. Once the problem is defined, the goals of intervention can be explicitly stated in terms of what should be happening when the goal is reached.

Goal-attainment scaling (Kiresuk and Sherman, 1968) offers a means of continuously evaluating how close a person comes to attaining the goal, or an organization comes to achieving the stated objectives in their contract with the service center. In most instances where preventive programs have been developed, a *contingency contract* between the group involved and the center should be explicitly stated. The contract would state the problems and goals, how and when they would be developed and attained, the consequences of reaching the goals and a means by which continuous scaled approximations to the stated goals could be recorded. Furthermore, the actual commitment to action of the contract, including fees, if applicable, could be attested to by the signatures of the participants. Problem-oriented flow charts, goal-attainment ratings, and contingency contracts provide a measurable and tangible basis for accounting for what is to be done, what has been done, and what is being done. And, of equal importance, the problem-oriented contract provides feedback into the process of how the program or intervention is progressing, whether or not the intervention involves individuals, groups or communities.

If the person, in the process of receiving treatment and aftercare, reduces his dependency on others and on society, then his outcome is one of improvement. If the *effort expended by society* increases, his outcome is negative. Effort expended can be rated in dollars and cents and it can also be rated by those who expend this effort (see Salmon, et al., 1966). If we are able to ascertain the effort needed to be expended and the cost or weighting of this effort, we could also define the goals of intervention in terms of effort or costs that must be alleviated.

Changes in the social behavior of society's members (changes in *community base rates*) also can be considered a concommitant of mental health or mental illness. However, these measures of change are subject to criticism as a function of not being able to hold all other environmental, social and ecological factors constant. It is assumed that the community will mirror the effects of the human service center.

In summary, a comprehensive system of delivering human services to all segments of the community has been discussed. Such a system, stressing positive mental health, would involve planning, prevention, treatment, maintenance and development of procedures for the perpetuation of these endeavors, as well as for their accountability and evaluation. Above all, such a system would provide better patient care or human services to individuals, to groups and to the community. Both the individual and society should benefit.

REFERENCES

Allport, G. W. *Pattern and Growth in Personality.* New York: Holt, Rinehart, and Winston, 1961.

Anthony, W. A., Buell, G. J., Sharratt, S., & Althoff, M. E. "Efficacy of psychiatric rehabilitation." *Psychological Bulletin,* 1972, 78, 447-456.

Atthowe, J. M., Jr. "Behavior innovation and persistence." *American Psychologist,* 1973a, 23, 34-41.

Atthowe, J. M., Jr. "Token economies come of age." *Behavior Therapy,* 1973b., In press.

Bandura, A. *Principles of Behavior Modification.* New York: Holt, Rinehart, and Winston, 1969.

Bergin, A. E. "The evaluation of therapeutic outcomes." In A. E. Bergin and S. L. Garfield (Eds.), *Handbook of Psychotherapy and Behavior Change: An Empirical Analysis.* New York: Wiley, 1971.

Cowen, E. L. "Social and community interventions." In Mussen, P. H. and Rosenzweig, M. R. (Eds.), *Annual Review of Psychology.* Palo Alto: Annual Reviews, Inc., 1973.

Cumming, E., & Cumming, J. *Closed Ranks: An Experiment in Mental Health.* Cambridge: Harvard University Press, 1957.

Dohrenwend, B. P. & Chin-Shong, E. "Social status and attitudes toward psychological disorder: The problem of tolerance of deviance." *American Sociological Review,* 1967, 32, 417-433.

Ehrlich, D., & Sabshin, M. "A study of sociotherapeutic oriented psychiatrists." *American Journal of Orthopsychiatry,* 1964, 34, 469-486.

Eysenck, H. J. "Learning theory and behavior therapy." *Journal of Mental Science,* 1959, 105, 61-75.

Eysenck, H. J. *The Effects of Psychotherapy.* New York: International Science Press, 1966.

Fairweather, G. W. (Ed.), *Social Psychology in Treating Mental Illness: An Experimental Approach.* New York: Wiley, 1964.

Fairweather, G. W., Sanders, D. H., Crissler, D. L. & Maynard, H. *Community Life for the Mentally Ill.* Chicago: Aldine, 1969.

Fromm, E. *The Art of Loving.* New York: Harper, 1956.

Glidewell, J. C., & Swallow, C. S. *The Prevalence of Maladjustment in Elementary Schools.* Chicago: University of Chicago Press, 1968.

Glueck, B. C., & Stroebel, C. F. "The computer and the clinical decision process: II." *American Journal of Psychiatry,* 1969, 125, 2-7.

Grant, R. L., & Maletzky, B. M. "Application of the Weed system to psychiatric records." *Psychiatry in Medicine,* 1972, 3, 119-129.

Gurin, G., Veroff, J., & Feld, S. *Americans View Their Mental Health: A Nationwide Survey.* New York: Basic Books, 1960.

Holden, C. "Nader on mental health centers: A movement that got bogged down." *Science,* 1972, 177, 413-415.

Houts, P. S., & Scott, R. A. *Goal Planning in Mental Health Rehabilitation.* Hershey, Pa.: The Milton S. Hershey Medical Center, 1972.

Johnson, W. G., & Pollack, I. W. "Efficiency and the Delivery of Mental Health Care." Unpublished paper, Rutgers Medical School, 1973.

Kiresuk, T. J., & Sherman, R. E. "Goal attainment scaling: A general method for evaluating comprehensive community mental health programs." *Community Mental Health Journal,* 1968, 4, 443-453.

Krasner, L. & Atthowe, J. M., Jr. "The token economy as a rehabilitative procedure in a mental hospital setting." In H. C. Rickard (Ed.), *Behavioral Interventions in Human Problems.* New York: Pergamon Press, 1971.

Krasner, L., & Krasner, M. "Token economies and other planned environments." In C. E. Thoresen (Ed.), *Behavior Modification in Education.* National Society for the Study of Education, 72nd Yearbook. Chicago: University of Chicago Press, 1973.

Krasner, L., & Ullmann, L. P. *Behavior Influence and Personality: The Social Matrix of Human Action.* New York: Holt, Rinehart, and Winston, 1973.

Leighton, D. C., Harding, J. S., Macklin, D. B., Macmillan, A. M., & Leighton, A. H. *The Character of Danger.* New York: Basic Books, 1963.

London, P. *The Modes and Morals of Psychotherapy.* New York: Holt, Rinehart, and Winston, 1964.

Lorion, R. P. "Socio-economic status and traditional treatment approaches reconsidered." *Psychological Bulletin,* 1973, 79, 263-270.

Maslow, A. H. *Motivation and Personality.* New York: Harper, 1954.

McDonough, J. M. "Making community health work: Organizing a work-for-pay program." *Psychological Reports,* 1973, 32, 127-134.

Mead, G. H. *Mind, Self and Society: From the Standpoint of a Social Behaviorist.* Chicago: University of Chicago Press, 1934.

Patterson, G. R. "Behavioral techniques based upon social learning: An additional base for developing behavior modification technologies." In Franks, C. M. (Ed.), *Behavior Therapy: Appraisal and Status.* New York: McGraw-Hill, 1969, Pp. 341-374.

Patterson, G. R. "Behavioral intervention procedures in the classroom and in the home." In A. E. Bergin and S. L. Garfield (Eds.), *Handbook of Psychotherapy and Behavior Change: An Empirical Analysis.* New York: Wiley, 1971.

Phillips, D. L. "Rejection of the mentally ill: The influence of behavior and sex." *American Sociological Review,* 1964, 29, 679-687.

Rabkin, J. G. "Opinions about mental illness: A review of the literature." *Psychological Bulletin,* 1972, 77, 153-171.

Rawls, J. *A Theory of Justice.* Cambridge: Harvard University Press, 1971.

Salmon, P., Atthowe, J. M., Jr., & Hallock, M. R. *A Method of Classifying Patients Receiving Long-Term Care.* San Mateo, Calif.: Department of Public Health and Welfare, 1966.

Sidman, M. "Operant techniques." In Bachrach, A. J. *Experimental Foundations of Clinical Psychology.* New York: Basic Books, 1962, Pp. 170-210.

Skinner, B. F. *Science and Human Behavior.* New York: MacMillan, 1953.

Skinner, B. F. *Beyond Freedom and Dignity.* New York: Knopf, 1971.

Srole, L., Langer, T. S., Michael, S. T., Opler, M. K., and Rennie, T. A. C. *Mental Health in the Metropolis,* Vol. I: *The Midtown Manhattan Study.* New York: McGraw-Hill, 1962.

Tharp, R. G., & Wetzel, R. J. *Behavior Modification in the Natural Environment.* New York: Academic Press, 1969.

Ullmann, L. P., & Krasner, L. *A Psychological Approach to Abnormal Behavior.* Englewood Cliffs, N.J.: Prentice-Hall, 1969.

Weed, L. L. *Medical Records, Medical Education, and Patient Care.* Cleveland: Case Western Reserve University Press, 1971.

Wexler, D. B. "Token and taboo: Behavior modification, token economies and the law." *California Law Review*, 1973, 61, 81-109.

Wilson, W. "Correlates of avowed happiness." *Psychological Bulletin*, 1967, 67, 294-306.

Wolpe, J. *Psychotherapy by Reciprocal Inhibition.* Stanford: Stanford University Press, 1958.

Yamamoto, K., & Dizney, H. F. "Rejection of the mentally ill: A study of attitudes of student teachers." *Journal of Counseling Psychology*, 1967, 14, 264-268.

Token economy systems are one of the most salient forms of behavioral treatment. The following chapter by Kazdin provides both an historical and futuristic overview of this treatment modality. In Kazdin's paper one can clearly see that behavior modifiers have come a long way toward becoming effective practitioners since the time when they were concerned almost exclusively with proving that the "law of effect" really works. It does! Now the crucial issues for behavioral treatment center around problems of response generalization, response maintenance, and the social and organizational parameters of treatment. Kazdin develops some interesting theories about the training of staff, and provides suggestions as to ways one can reinforce the reinforcers. In addition, his paper demonstrates how token economy treatment systems provide unique opportunities to monitor patient behaviors as the patient moves from institution to community life.

Kazdin's paper is an excellent review of research on token economy systems. His conclusions are optimistic, yet point out the difficulties surrounding research in general, and token economy programs in particular. This paper provides good examples of how one can use a rational, behavioral treatment program to generate not only behavioral changes in patients but also comprehensive changes in supporting social systems.

Alan Kazdin

CHAPTER 10

A Review of Token Economy Treatment Modalities

TOKEN ECONOMY PROGRAMS

In recent years, operant techniques have been applied extensively to individuals who exhibit a variety of deviant behaviors. Major impetus for this application in mental health settings was the systematic work of Ayllon and his colleagues with institutionalized psychiatric patients. In different reports, several operant techniques were used to train patients to feed themselves, to behave appropriately in dining facilities, to dress appropriately, and to cease engaging in bizarre habits, psychotic talk, and somatic complaints (Ayllon, 1963, 1965; Ayllon & Azrin, 1964; Ayllon & Haughton, 1962, 1964; Ayllon & Michael, 1959). (See Davison, 1969 for a review of this work.) Other earlier applications of operant principles have also demonstrated the efficacy of various techniques in individual cases (Ullmann & Krasner, 1965; Davison, 1969; Eysenck, 1964).

There are several features which characterized many of the early applications of operant principles. Frequently the application of operant techniques was restricted to only a few individuals at the same time (Issacs, Thomas, & Goldiamond, 1960). Procedures were often carried out during experimental sessions rather than during the ordinary ward routine (King, Armitage, & Tilton, 1960). Furthermore, only one or a few reinforcers were used (Azrin & Lindsley, 1956) while focus was restricted to only a few target behaviors at one time (Wickes, 1958). Many applications were of a demonstrational rather than functional nature (Ferster & DeMyer, 1961). Finally, careful analysis of response characteristics sometimes served as a major goal (Barrett & Lindsley, 1962; Lindsley, 1960).

Of course, there are early studies which do not include these "characteristics". For example, Ayllon and Haughton (1962) included 32 psychiatric patients with feeding problems, in a program designed to alter mealtime behavior. This was hardly an attempt to treat one case, in experimental sessions, with a demonstrational rather than a therapeutic goal. While there are other exceptions as well, most early studies have been restricted in focus.

Along parallel lines, other advances were being made in the application of operant principles to human subjects. Ferster and DeMyer (1961, 1962) and Staats and his associates (Staats, 1968; Staats & Butterfield, 1965; Staats, Minke, Finley, Wolf, & Brooks, 1964; Staats, Staats, Schultz, & Wolf, 1962) were investigating the use of generalized conditioned reinforcers (Kelleher & Gollub, 1962). In the Staats' research, children were trained to perform various tasks, such as reading, for long periods of time. Tokens earned for performance on the tasks were exchangeable for a variety of back-up rewards. Several additional studies have contributed greatly to the present form of reinforcement practices in treatment and rehabilitation settings by providing carefully executed functional analyses of individual cases.

Token economies have stemmed directly from much of the work outlined above (cf. Ayllon & Azrin, 1968; Kelleher & Gollub, 1962). Token economies extend previous practices by incorporating many individuals into the contingencies, focusing on behavior on the ward, employing a variety of reinforcers, consequating several behaviors, and serving primarily therapeutic goals. Token economies have been implemented in numerous settings with diverse populations (Carlson, Hersen, & Eisler, 1972; Kazdin & Bootzin, 1972; Liberman, 1968; Milby, 1972; Turton & Gathercole, 1972).

Rather than a review of past token economies, the present discussion will focus on recent advances and apparent trends in this area. As the demonstrations reported in early token programs were replicated, the goals of many programs became increasingly ambitious and sophisticated. Several recent trends demonstrate this, including: focus on "symptomatic" behaviors and social skills, in addition to routine ward behavior; comparison of token programs with other procedures; assessment of generalized effects produced by token reinforcement beyond changes in specific target responses; evaluation of the role of punishment; and evaluation of economic principles which may complement reinforcement principles in predicting client behavior.

Token programs have positive implications for the delivery of mental health services. In this chapter, we will discuss these implications, and their relation to the changing role of patients, staff, and the community at large.

"Symptomatic Behaviors" and Social Interaction of Psychiatric Patients

Numerous studies have focused upon general ward behavior of patients. Included in these are self-care skills, grooming, attending and participating in activities, engaging in jobs on and off the ward, and similar behaviors adaptive in the hospital (Allen & Magaro, 1971; Arann & Horner, 1972; Atthowe & Krasner, 1968; Ayllon & Azrin, 1965; Cohen, Florin, Grusche, Meyer-Osterkamp, & Sell, 1972; Hartlage, 1970; Heap, Boblitt, Moore, & Hord, 1970; Lloyd & Garlington, 1968; McReynolds & Coleman, 1972; Schaefer & Martin, 1966; Steffy, Hart, Craw, Torney, & Marlett, 1969; Suchotliff, Greaves, Stecker, & Berke, 1970; Winkler, 1970.) These show that there is little question of the importance of developing adaptive behaviors in institutionalized patients, particularly since institutionalization often fosters maladaptive behaviors (Paul, 1969).

Initially, token economies (e.g., Ayllon & Azrin, 1965) focused upon non-symptomatic (as opposed to idiosyncratic) behaviors to establish the amenability of general ward behaviors to token reinforcement, and to displace symptoms with functional behaviors. However, as this use of token economies progressed it became apparent that bizarre behaviors could be altered by establishing functional behaviors. For example, O'Brien and Azrin (1972) used tokens to alter the high rate of screaming of a female schizophrenic patient. Tokens were delivered for "functional behaviors" such as social skills, housekeeping, and grooming (which were not necessarily incompatible with screaming). The increase in positive behaviors with token reinforcement resulted in a decrease in screaming. The authors suggest that positive behaviors "functionally displaced" the screaming.

Several investigators have focused directly upon target "symptoms" rather than displacing them by increasing other behaviors. The direct treatment of "symptomatic behaviors" represents a desirable trend in token programs. While altering bizarre behaviors in patients is not new (e.g., Ayllon, 1963; Ayllon & Michael, 1959), the focus on those bizarre behaviors (symptoms) which have led to particular diagnoses do present important information as to the mutability of relatively complex behaviors. The thrust of token programs will be most apparent as these latter responses continue to be altered.

Wincze, Leitenberg, and Agras (1972) found that token reinforcement effectively altered delusional talk in therapy sessions with paranoid schizophrenics. Feedback for delusional talk (i.e., "that was incorrect") did not consistently alter verbal behavior. Evaluation of verbal behaviors on the ward showed less dramatic change even though tokens were delivered for nondelusional talk with the staff.

Kazdin (1971a) suppressed delusional psychotic speech in an adult pre-psychotic retardate by contingently withdrawing tokens. The statements decreased dramatically and did not recover over a six-month follow-up period. Similarly, Bartlett, Ora, Brown and Butler (1971) reduced psychotic speech in an autistic child and developed rational speech by using token reinforcement.

Reisinger (1972) treated "anxiety depression" in an institutionalized patient (diagnosis not specified) with tokens that were delivered for smiling, and withdrawn for crying. At the end of treatment the tokens were eliminated and social reinforcement alone was delivered for smiling while crying was ignored. This effectively maintained smiling for the four remaining weeks before the patient was discharged.

An equally significant response class for psychiatric patients is the development of social behaviors. Of course, a precondition for social interaction may be elimination of certain bizarre behaviors. However, this alone will not insure the appearance of social behaviors. Liberman (1972) trained four withdrawn and verbally inactive psychiatric patients to engage in conversation with a group. In one experimental phase, a group contingency was used whereby tokens earned for total number of conversation units within the group were equally divided among the members. Individualized reinforcement subsequently replaced the group contingency. Conversation increased dramatically under both types of contingencies.

Similarly, Bennett and Maley (1973) reinforced psychiatric patients for conversing with each other in experimental sessions. Social interaction increased in the sessions and on the ward as well. In another study, Kazdin and Polster (1973) developed social interaction in two adult retardates who were severely withdrawn.

By providing token reinforcement for conversation, Leitenberg, Wincze, Butz, Callahan and Agras (1970) treated a patient who avoided social interaction. However, instructions without reinforcement did not alter behavior. The patient proved to have had an additional problem, fear of injury. Performances of several behaviors related to the fear were reinforced with praise. This regimen dramatically increased time spent in the feared activity.

Behaviors focused upon in token programs have extended well beyond grooming and self-care skills. Routine ward behaviors are still exceedingly important, particularly since reinforcement of these behaviors is sometimes associated with generalized effects (Atthowe & Krasner, 1968; Maley, Feldman, & Ruskin, 1973). However, development of social skills, appropriate interaction patterns, and especially counter-symptom behaviors is likely to be crucial for community adjustment (Freeman & Simmons, 1963).

Concomitant Changes Following Token Reinforcement

Initial studies on token reinforcement evaluated the effect of the contingencies solely on the target behaviors — possible additional changes resulting from the contingencies were never considered. In recent years, however, there has been an added interest, viz., evaluation of effects which extend beyond changes in the target, or reinforced, behaviors (cf., Kazdin, 1973b, 1973c), specifically in the area of *response generalization.* Several studies have revealed that changes in the frequency of a response are associated with changes in topographical features such as intensity or quality of the response (Burchard & Tyler, 1965; Hauserman, Zweback & Plotkin, 1972; Hawkins, Peterson, Schweid & Bijou, 1966).

Shean and Zeidberg (1971) compared token reinforcement procedures with traditional care in developing self-care and role behaviors, and increasing extra-hospital visits in psychiatric patients. Token reinforcement led to greater changes in ratings of cooperation on the ward, communication skills, social interaction, participation in hospital activties, time out of the hospital, and a reduction in the use of medication, relative to the control ward.

The majority of these improvements, however, were not directly shaped with reinforcement. Winkler (1970) used token reinforcement and response cost to alter behaviors of chronic psychiatric patients to develop self-help skills, participation in exercises, and attendance to meals. Violence and noise on the ward, which were initially excluded from the contingencies, decreased while the token program was in effect.

Maley et al. (1973) compared psychiatric patients reinforced with tokens for several behaviors on the ward, using controls who received typical custodial treatment. Dependent measures included several behaviors found in a standardized interview and ratings of video-taped behavior samples. Patients in the token economy ward were found to be better oriented, and more able to perform a discrimination task, to follow complex commands, and to handle money in making business purchases than control patients. Moreover, ratings of behavior indicated that token economy patients showed more appropriate mood states, were more cooperative, and displayed better communication skills than control patients.

Importantly, in later rating, token economy patients needed less hospitalization and were more likeable. Since patients were reinforced for specific behaviors on the ward, (e.g., grooming) the beneficial effects obtained for behaviors not included in the contingencies represent generalization. Similarly, Bennett and Maley (1973) noted changes in mood, communication, and social skills in two psychotic patients who were reinforced for talking with each other. Moreover, reinforced patients showed greater token earnings on the ward and greater participation in activities than controls. Since the behaviors specifically reinforced in the experimental sessions were restricted verbal interactions, the effect of the contingencies generalized considerably.

Gripp and Magaro (1971) evaluated a token economy with schizophrenic patients. Patients received tokens for job performance and other "pre-selected desirable behaviors". Numerous measures were used to evaluate the effect of token reinforcement after six months of the program. Patients on the token economy ward showed significant improvement in withdrawal, thought disorder, and depression, on the Psychotic Reaction Profile. Nurses rated patients as improved in social competence, neatness, irritability, and manifest psychosis. Improvement in these areas was either markedly less or not apparent on control wards which did not receive token reinforcement. It is difficult to discern whether the reported changes are evidence of response generalization, since the behaviors which were reinforced were not made explicit. Nevertheless, this study represents a trend toward looking for general effects of token programs.

Mulligan, Kaplan, and Reppucci (1971) evaluated changes in "cognitive" variables resulting from token reinforcement delivered to elementary school students. Tokens were delivered for appropriate classroom behavior and completion of tasks. As a result of the program, gains were reported in I.Q., and arithmetic achievement scores. A slight decrease was also noted in anxiety. Kubany, Weiss, and Sloggett (1971) altered disruptive classroom behavior of a six-year old boy. Punctual return to class from recess improved while the contingencies were in effect without reinforcement for this behavior. Similarly, Twardosz and Sajwaj (1972) reported that token reinforcement for in-seat behavior of a hyperactive child increased appropriate social interaction with peers as well as individual play behaviors.

Horton (1970) demonstrated generalization of responses with delinquent boys in a home for emotionally disturbed children. In a reversal design, tokens (backed by money) were administered for aggressive responses (hand slapping of a peer) on one task. Aggressiveness on another task was monitored throughout the experiment. The effect of reinforcement for aggressive responses on one task generalized to other forms of aggressive responses on a different task.

In spite of favorable evidence for generalization of effects to responses not directly reinforced, this does not always occur. For example, Ferritor, Bucholdt, Hamblin, and Smith (1972) evaluated the effect of token reinforcement on attentive behavior and correct academic performance in an elementary school classroom. Although reinforcement for attentive behavior altered this behavior, academic work did not improve. Conversely, when reinforcement was given for correct completion of academic work, attentive behavior did not improve. Both behaviors improved only when they were both reinforced.

The authors caution against hoping for by-products of reinforcement contingencies. If a particular behavior change is required, the behavior should be included into the contingencies. While these recommendations are well taken, the increasing evidence for generalization across responses in several areas of behavior therapy is rather compelling (e.g., Bandura, Blanchard & Ritter, 1969; Kazdin, 1973b; Meichenbaum, 1969; Paul, 1967). Concomitant behaviors should be assessed to determine whether additional changes are associated with those effected in target response areas.

Comparative Studies and Combined Treatments

The majority of token economy programs have been evaluated with intra-subject replication designs in the absence of comparative groups (Kazdin & Bootzin, 1972). Of course, several "own control" designs can be used to evaluate a program (Kazdin, 1973e). However, many questions require the use of comparison groups. Salient questions include: the relative efficacy of token programs to traditional procedures and to untreated groups, and the magnitude of behavior change attributable to the token program per se. Moreover, investigation of the contribution of specific variables or parametric variations of a given variable in a token economy require inter-group comparisons to avoid confusing the effects due to sequence of experimental conditions or multiple treatment interference (Kazdin, 1973e). An increasing number of studies have compared token economy procedures to other treatments including, but not limited to, custodial ward care. Similarly, investigators are beginning to combine token reinforcement with other treatment strategies.

Marks, Sonoda, and Schalock (1968) compared reinforcement and relationship therapy with chronic schizophrenics. Patients were divided into two groups. Each group received both treatments but in a different order. Relationship therapy involved individual psychotherapy. Reinforcement consisted of tokens delivered by staff (at their discretion) for improvement in individualized areas of problems (e.g., personal appearance, expressing feelings). On several measures there were no differences between treatments. Both methods were concluded to be effective. The authors note the operational difficulties in keeping the treatments distinct (e.g., avoiding reinforcement in relationship therapy). Since the reinforcement program was not evaluated on specific target behaviors, and reinforcement was not delivered in a consistent fashion for well specified responses, the results are difficult to interpret.

Hartlage (1970) compared traditional therapy (insight) and reinforcement therapy (social reinforcement, privileges, and consumables) for "adaptive" responses. Comparisons of pre- and post-treatment adjustment showed greater adjustment resulting from reinforcement therapy. On self-concept data, treatments were not significantly different. Therapists' ratings of improvement favored reinforcement therapy.

Birky, Chambliss, and Wasden (1971) compared the efficacy of token economy versus traditional ward care on discharge rates. Although there were no differences in the number of patients discharged from the token economy and two control wards (i.e., patients who remained out of hospital for at least six months), the patients discharged on a token economy ward had a significantly greater length of previous hospitalization. Thus, the reinforcement procedure had greater effect in discharging patients who were more chronic.

Gripp and Magaro (1971) made rather extensive comparisons of token reinforcement with routine hospital treatment. Schizophrenic patients on a token economy ward made several changes on scales reflecting cognitive and affective concomitants of psychoses; whereas, fewer gains (and some regression) were noted for patients on control wards.

One of the best comparative studies methodologically, was recently reported by Olson and Greenberg (1972). Psychiatric patients were exposed to one of three treatments for four months: milieu therapy, interaction (milieu plus two hours of weekly group therapy), and token reinforcement. The reinforcement group received tokens for making group decisions regarding ward administration and for attendance to activities. Patients in the token reinforcement condition were superior to other groups in the areas of number of patients spending days on town passes, days out of the program, and attending activities. Although not all measures favored the token system (e.g., ratings of social adjustment), the program tended to be more effective than other procedures.

Heap et al. (1970) compared behavior-milieu therapy consisting of token reinforcement, ward government, and other adjunctive procedures, with traditional ward care in developing self-care skills. The reinforcement group showed greater performance of self-care skills and rate of discharge than the control ward. However, the conclusions of this study have been criticized for several methodological flaws (Carlson et al., 1972; Kazdin, 1973e).

Comparative studies of the effectiveness of token programs require extensive methodological effort on the part of the investigators. Frequently, the requirements for adequate experimental design are not easily met. Understandably, several methodological problems delimit the clarity of the results. Although token programs seem to result in greater therapeutic effects relative to other procedures in psychiatric hospitals, the program may be confounded with a change of environment (Heap et al., 1970), selection of special staff for the token ward (Gripp & Magaro, 1971), initial differences in patients (Shean & Zeidberg, 1971), and expectation for behavior change (Gripp & Magaro, 1971).

A potential problem in adding token reinforcement to other existing therapy procedures is that the combination is not evaluated empirically. It is unclear whether the combined treatment is any better than the component procedures given alone. Hauserman et al. (1972) used token reinforcement to increase initiation of verbal statements of hospitalized adolescents participating in group therapy. In a reversal and multiple-baseline design, the effect of tokens was demonstrated. However, the contribution of group therapy to behavior change was not evaluated. Carpenter and Caron (1968) used tokens (S & H Green stamps) to reinforce pre-adolescent delinquent boys in group meetings. Target behaviors included attendance, talking during discussions, punctuality, improvement in grades, and other behaviors. Additionally, psychodynamic techniques such as catharsis, ego support, and others were used in the group. Since no data are presented, the contribution of the combined procedure to behavior change is unclear. While there may be several advantages in combining token reinforcement with other procedures, it would be useful to determine whether the other procedure or the token program alone is less effective than the combination.

Role of Punishment in Token Economies

Token economies attempt to rely heavily on the use of positive reinforcement of appropriate behaviors even when the goal is to decrease inappropriate behavior. Despite this emphasis, an increasingly large number of token programs employ some form of aversive control. Whereas token reinforcement contingencies have been subjected to careful scrutiny, punishment has not been systematically evaluated in a number of programs (Kazdin, 1972b). Usually two forms of punishment, either alone or in combination, are used. They are reinforcer withdrawal, in the form of time-out from reinforcement, or response cost. The distinction between time-out and response cost is that in time-out there is a period of *time* during which reinforcers are no longer available (Leitenberg, 1965), whereas with response cost, there is no necessary temporal restriction for earning further reinforcers (Kazdin, 1972b).

Isolation typifies time-out, whereas a fine (loss of tokens) or loss of privileges typify response cost. The distinction becomes blurred in some token programs where clients are punished by having to *pay* more *tokens* for back-up reinforcers than the usual cost for a certain *period* of *time* (Winkler, 1971). Indeed, some authors have suggested a synthesis in conceptualizing time-out and response cost (e.g., Striefel, 1972). Of course, time-out and response cost have been investigated independently of the token economy (Sherman & Baer, 1969; Kazdin, 1972b). However, the use of these procedures in token economies has not been exhaustively explored.

Recent investigations exemplify the use of punishment in token programs. Burchard and Barrera (1972) evaluated response cost (fines) and time-out (isolation) in a token economy for mildly retarded antisocial boys. Punishment was used to suppress swearing, personal assault, damaging property, and similar acts. Four experimental groups were exposed to different time-out/response-cost combinations. Variations were made in the amount of time of isolation (from 0 minutes to 30 minutes), and the amount of the fine (from 0 tokens to 20 tokens). Larger fines and longer time-out periods led to greater response suppression. Unexpectedly, both low-magnitude time-out durations and fines led to an increase in instances of deviant acts.

Kaufman and O'Leary (1972) demonstrated that token loss for inappropriate behavior and reinforcement for appropriate classroom behavior were equally as effective in altering behavior. This suggests that response cost alone can be used effectively as a behavior modification technique.

Winkler (1970) showed that fines were effective in controlling episodes of violence and noise. Although these behaviors had changed as a result of the generalized effects of other contingencies, levying a cost was effective in making further reductions. When fines were lifted, the disruptive behaviors increased.

Upper (1971) used a ticket system to reduce rule violations on a psychiatric ward. Whenever a patient committed an infraction, a ticket was administered. Subsequently, tokens were subtracted from the amount normally received at the end of the day. The data suggest the efficacy of the cost procedure, although no reversal phases were included in the design.

In another study, Parrino, George, and Daniels (1971) charged patients tokens for taking pills (those which seemed inessential). By the end of a twenty-week period, there was a substantial reduction in the use of medication. However, "punishment" procedures are not always effective. Boren and Colman (1970) found that response cost with hospitalized delinquent soldiers was ineffective in increasing attendance to an activity. Token fines were levied for staying in bed instead of attending a group meeting. When fines were invoked, attendance became *worse*.

Based upon laboratory evidence, Azrin and Holz (1966) concluded that punishment can be efficiently used by combining it with reinforcement for alternate responses. Although many programs use reinforcement and punishment, different, and seemingly unrelated, behaviors may be under each type of contingency. The beneficial effects of combining each procedure may be maximized by delivering reinforcers for a response which is incompatible with, or functionally displaces, the response to be suppressed. The combined use of punishment and reinforcement is easily implemented when the occurrence of a behavior is reinforced, and its nonoccurrence is punished (or vice versa), along the same reinforcer dimension (e.g., token presentation and withdrawal). For example, Kubany et al. (1971) combined token reinforcement and time-out to control behavior of a disruptive boy. Accumulated time on a clock earned stars and back-up reinforcers (shared by peers). Time-out consisted of terminating the running clock when disruptive behavior was performed. As mentioned earlier, Reisinger (1972) combined response cost and reinforcement to alter "anxiety" depression of an institutionalized patient. Token and social reinforcement for smiling were paired with response cost for crying. Ultimately, smiling was maintained with social reinforcement.

The use of punishment in applied settings (Johnston, 1972), requires the evaluation of several issues; such as, the presence of any undesirable side effects, the permanence of response suppression, and the parameters which contribute to its efficacy. As attempts to alter more complex behavior increase, the use of punishment may increase.

Role of Economics in Token Economies

Until a few years ago token economies were explained, at least by behavior modifiers, almost entirely in terms of operant principles or reinforcement theory. Any limitations of the procedures were considered to reflect, in part, failure to implement the procedures adequately (Kazdin, 1973d). Yet the recent work of Winkler and colleagues (Ingham, Andrews, & Winkler, 1973; Kagel & Winkler, 1972; Winkler, 1971, 1972; Winkler & Krasner, 1972) shows that patient performance in a token economy depends upon more than the parameters of reinforcement. Specifically, several notions from economic theory can supplement reinforcement theory and make specific predictions regarding patient performance.

First, the notion of *consumption schedule* is relevant to patient behavior in a token economy. This schedule describes the relationship between income and expenditures. Individuals with more income spend more. Generally, spending falls within ten percent of income. Individuals who have extremely low incomes spend more than this, and individuals with high incomes spend far below this percentage. A problem with extrapolating the relationship from national economies to token systems is that certain conditions are not always met in token programs.

For example, for a person who earns few tokens to spend more than his earnings requires that he either borrow tokens or use an accumulated savings (Fetke, 1972). Yet, in most token economies borrowing is not allowed because this might constitute non-contingent reinforcement if not carefully executed. In addition, if accumulated savings are used, performance can be independent of income altogether. Nevertheless, the general relationship following consumption schedules holds (Winkler, 1971). At high levels of income, spending attenuates, and presumably tokens take on less value. Winkler (1972) has also noted that the discrepancy between income and expenditure, or *savings* in token economies may reach a *critical limit.* If savings fall below an individual's critical limit, token-earning behaviors should increase; if savings fall within a critical range, performance should reach an equilibrium (i.e., no change in performance). Of course, the goal of most token economies is to increase performance, not to stabilize performance at low rates.

The implications of consumption schedules for token economies are clear. If token-earning behaviors and token income are to increase one of several solutions can be used: 1) prices of back-up reinforcers can be increased, 2) the value of tokens can be decreased, thereby altering the value of savings, and 3) the range of goods (back-up reinforcers) available can be increased. Any of these solutions can be used to alter the level of earnings. As Winkler (1972) has noted, if reinforcement does not affect consumption schedules or critical range of savings, performance will stabilize levels of performance and earnings.

A second principle from economics is referred to as *Engel's Law.* Simply stated, as income increases, the proportion of total expenditure spent on urgent needs decreases, whereas, expenditure on luxuries increases. Indeed, this relationship was supported in token economy expenditures when meals were considered essential and canteen items were considered luxuries (Winkler, 1971). One implication is that items which can be considered luxuries should be expanded because they will be in increased demand as incomes increase.

A third principle relates to the responsiveness of consumer demand to price changes. The economic notion of elasticity of a demand curve measures this responsiveness, and is defined as the percentage change in demand resulting from a one percent change in price. For some items, a price drop leads to a relatively larger percent demand than the actual drop (i.e., items with elastic demands). That is, expenditures increase. For other items, a decrease in price will result in an increase in demand that is smaller than the price drop (i.e., items with inelastic demands). Necessities usually are inelastic, whereas, luxury items are usually elastic. On an applied level, if prices of back-up reinforcers are to be increased in a token economy, items with inelastic demands (necessities) should be made expensive to stimulate spending. If the price of items with elastic demands (luxuries) are increased, spending may decrease. It may be useful to catalog the relative elasticity or inelasticity of items which can best be used to stimulate spending, particularly for individuals who respond minimally to the program (Kazdin, 1973d).

Operant principles do not always make clear predictions about expenditures, earnings, and performance. A few concepts are employed, although their relation to performance is not precise enough to be easily tested. However, economic principles can be used to supplement reinforcement theory in clarifying the relationship between performance and increased token earnings. For example, it is clear that magnitude of token reinforcement can be increased to stimulate performance in token economies (Wolf, Giles & Hall, 1968). Satiation is considered to be one limiting condition for unlimited gains derived from increases in magnitude. It is likely to occur when token savings begin to surpass the critical range of savings. Since there are means to estimate an individual's critical level of savings (Winkler, 1972), predictions as to *when* satiation will begin can be advanced. More importantly, satiation can be abated by altering the prices of back-up items or the value of tokens, reducing savings, or increasing the range of available back-up items. These latter solutions are predicted from economic theory.

RESPONSE MAINTENANCE AND TRANSFER OF TRAINING

Since the efficacy of token programs has been firmly demonstrated, the major issue becomes the achievement of long-term maintenance of responses following program termination, and transfer of these responses to extra-treatment settings. Several reports have indicated improved discharge and/or re-admission rates following participation in a token economy (Atthowe & Krasner, 1968; Birky et al., 1971; Ellsworth, 1969; Heap et al., 1970; Henderson & Scoles, 1970; Stayer & Jones, 1969). However, the precise role of the reinforcement contingencies in these measures is not always clear (Kazdin & Bootzin, 1972). There is considerable evidence that responses developed with token reinforcement may not be maintained once the contingencies are withdrawn, and are not likely to transfer automatically to nontreatment settings. For this reason, the areas of response maintenance and transfer of training remain the most fertile for research.

It should be noted at the outset, however, that behaviors developed through token reinforcement do not always return to baseline levels after the contingencies are withdrawn (Hewett, Taylor & Artuso, 1968; Kazdin, 1971a; Kazdin, 1973f; Medland & Stachnik, 1972; Surratt, Ulrich & Hawkins, 1969; Whitman, Mercurio & Caponigri, 1970), although these are exceptions (Kazdin & Bootzin, 1972). When behaviors are maintained, the reason is frequently unclear. There are several interpretations proffered to account for unplanned response maintenance.

First, it is possible that behaviors developed through token reinforcement come under control of extra-experimental reinforcers (Baer, Wolf, & Risley, 1968; Bijou, Peterson, Harris, Allen, & Johnston, 1969). Events associated with tokens may acquire conditioned reinforcement value and serve to maintain behaviors following token withdrawal (Medland & Stachnik, 1972). For example, investigators working in a classroom setting have suggested that teachers may more readily function as secondary reinforcers after being associated with a token economy than before (Chadwick & Day, 1971). A second interpretation of unplanned maintenance is that after tokens are withdrawn other reinforcers which result directly from the activities themselves maintain behavior. For example, token reinforcement for social interaction may be unnecessary to maintain behavior because of the "natural" reinforcement which follows (but see Kazdin & Polster, 1973). A third interpretation is that a token program alters staff behavior in some permanent fashion so that they continue desirable response consequation after tokens are discontinued. Even though tokens are withdrawn, staff are utilizing contingent reinforcement, prompting, and executing other procedures developed in their repertoires during the token program. In spite of the reasonable nature of each of these explanations, and partial support which might be provided for each, it is still usually a matter of conjecture as to why a response is maintained without specific programming.

Programming response maintenance is usually required since removal of the contingencies frequently results in a decrease in appropriate performance. Several procedures which might be useful in achieving maintenance have been detailed elsewhere (Kazdin & Bootzin, 1972). The procedures are not necessarily independent, mutually exclusive, or exhaustive. The list of procedures used to enhance response maintenance includes: 1) Systematically substituting social reinforcers for the tokens. Praise alone is effective in maintaining behavior following token reinforcement (Chadwick & Day, 1971; Reisinger, 1972; Wahler, 1968). 2) Gradually fading tokens after consistent and prolonged performance of the target behavior. If behavior has been performed consistently over a long period, there may be little or no dependence upon tokens (e.g., Atthowe & Krasner, 1968; Garlington & Lloyd, 1966; Lloyd & Abel, 1970; Schaefer & Martin, 1969). The use of levels or steps through which patients progress represents one way in which tokens may be systematically faded (Kazdin & Bootzin, 1972). 3) Exposing the individual to extra-treatment settings while tokens are being faded (Kelley & Henderson, 1971). Presumably, this procedure helps bring appropriate behaviors under control of the stimuli under which the individual will ultimately function. 4) Training individuals such as parents, teachers, and peers found in the client's environment, to continue contingencies (e.g., Henderson & Scoles, 1970). Since target behaviors are often readily altered in the natural environment, paraprofessionals can be trained to carry out the entire program (see Ayllon & Wright, 1972 for review). When the natural environment is the therapeutic environment, behaviors may be more likely to be maintained than when training takes place in a treatment facility, because of the discriminations which may be made between treatment and extra-treatment settings (Kazdin, 1971b). 5) Scheduling reinforcement to build resistance to extinction. When behavior is reinforced on a very thin schedule (Phillips, Phillips, Fixsen, & Wolf, 1971), or when reinforcement is withdrawn entirely (Kazdin & Polster, 1973), behavior may be maintained. It is unclear whether intermittent reinforcement will result in transient resistance to extinction or long-term maintenance. 6) Self-reinforcement or self-regulation (Kanfer, 1970) might result in response maintenance. If the individual can be trained to monitor and provide consequences for his own behavior, behavior may be maintained in a variety of settings and situations. Advantages of self-reinforcement over externally regulated token reinforcement have been suggested with children (Glynn, 1970; Lovitt & Curtiss, 1969). However, whether self-reinforcement training results in greater resistance to extinction when reinforcement is terminated is equivocal (Bolstad & Johnson, 1972; Johnson, 1969, 1970). 7) Training in self-instruction is related to the self-reinforcement procedure outlined above but focuses on antecedent events which the client can use to control his own behavior. Meichenbaum (1969) noted a serendipitous finding that self-instruction facilitated generalization of trained responses across tasks, time, and situations. Subsequent experiments demonstrated that altering the instructions one gives to oneself exerts control over a variety of nonverbal behaviors (Meichenbaum & Cameron, 1973; Meichenbaum & Goodman, 1971). Importantly, many behaviors altered with self-instruction seem to be maintained over time. 8) Manipulating reinforcement delay might also enhance resistance to extinction. Two variations are possible: varying the delay between performance of the target response and token delivery (Atthowe & Krasner, 1968), and increasing the delay between token delivery and exchange of tokens for back-up rewards (Cotler et al., 1972). 9) Manipulating several parameters of reinforcement simultaneously (i.e., varied reinforcement) might be useful in augmenting resistance to extinction. A number of aspects of reinforcement such as magnitude, delay, place, quality, and schedule can be altered. Laboratory evidence suggests that the greater the number of sources of variation, the greater the resistance to extinction (McNamara & Wike, 1958).

Although each of these methods might be useful in response maintenance, there has been little systematic use of these procedures in token economies (Kazdin & Bootzin, 1972). The effect of specific procedures following reinforcement withdrawal has been dealt with in a few studies examing schedules of reinforcement. Phillips et al. (1971), provided token reinforcement and fines to control room-cleaning behavior. In a final experimental phase, point consequences were faded using an "adjusting consequence" procedure. The number of days in which room cleaning was checked was gradually decreased (i.e., increased intermittency). However, on any day that the response was checked, the number of points earned or lost was multiplied by the number of previous days in which behavior had not been checked. This procedure ensured that both the frequency and magnitude of reinforcement were not reduced simultaneously, since the potential number of points remained the same during the fading period. Ultimately, in this report, performance remained high even though checks were made only eight percent of the time.

Kazdin and Polster (1973) evaluated token reinforcement on the behavior of two withdrawn adult retardates. Continuous token reinforcement for social interaction led to marked increase in this behavior, but the effects were lost in a reversal phase. Subsequently, one client was given continuous, and the other, intermittent reinforcement. In a final extinction period, the client who previously had been reinforced intermittently maintained a high rate of social interaction for the five week follow-up period. The continuously reinforced subject, however, returned to base levels of social interaction. These findings suggest that intermittent schedules may prove useful in delaying extinction.

A great deal of additional work is required to determine what procedures can be used to maintain behavior. One solution, of course, is to substitute other behavior modification programs after token reinforcement is withdrawn (e.g., Walker & Buckley, 1972). However, in many settings where programs are not continued in any form, resistance to extinction must be enhanced with little or no aid from a specifically designed and carefully executed program in the new environment.

Many of the above comments are restricted to building resistance to extinction. Response maintenance is often a prior issue to stimulus generalization or transfer of behavior to nontreatment settings. A patient's behavior needs to be maintained in the hospital when the program is withdrawn before considering transfer of the behavior to a non-hospital setting. (However, a concern might be whether target behaviors transfer to situations in which the contingencies are not being carried out while the program is in effect in one setting.)

Generalization of behavior across situations seems to be the exception (Bennett & Maley, 1973; Kazdin, 1973f; Walker, Mattson, & Buckley 1971) rather than the rule (see Kazdin & Bootzin, 1972 for review). Several of the procedures outlined above might be useful to enhance generalization across situations. In particular, training individuals in the nontreatment environment to administer the contingencies, developing self-reinforcement and self-instructional skills, and providing reinforcement in several settings during the program should facilitate transfer of training. It is desirable to provide reinforcement in a variety of settings using a number of agents to administer the contingencies so that there is as broad a discriminative stimulus control over behavior as possible.

Probably there are several additional procedures which may be useful in building response maintenance and insuring transfer of training to extra-treatment settings. Effective maintenance and transfer strategies need to be isolated so that they can be readily incorporated into ongoing programs. Ultimately, the success of token programs will be assessed by how well behaviors are maintained outside of the training setting rather than the extent of behavior changes occurring in the training setting when the contingencies are in effect.

SOCIAL AND ORGANIZATIONAL IMPLICATIONS OF TOKEN ECONOMIES

The delivery of mental health services through token reinforcement programs has led to new roles for patients, paraprofessionals, and the community. The manner in which treatment is delivered, the participation of clients in their own treatment, the therapeutic contribution of staff, and relation of treatment efficacy to community participation represent interesting social and organizational concomitants resulting from operant programs.

Use of Peers and Patients in Token Economies

Patients participating in a reinforcement program often assume a great deal of responsibility for their own behavior, the behavior of their peers, and the administration of the token economy itself. Increasingly, peers (fellow patients, clients, residents, and students) are employed in some aspect of the token economy either to serve in a function that is usually reserved for staff (e.g., data collection, token delivery), or to augment the efficacy of the contingencies. A direct attempt is made to replace dependency, apathy, and institutionalization with responsibility and active participation (Atthowe & Krasner, 1968; Schaefer & Martin, 1966). Instead of having treatment imposed upon them, patients are called upon to make decisions about matters related to treatment (Pomerleau et al., 1972). The decisions are not just "token" acts to convey a quasi-democratic structure of the institution but require serious planning and decision making on the part of patients. The expanded role of the patient is clear in token programs where patients can participate in their own "treatment" (e.g., deciding policy in the program or designing contingencies) and in the "treatment" of others (e.g., collecting and summarizing program data) (Kazdin, 1972a).

Patients may be directly employed in activities related to running and evaluating the token program. Dominguez, Acosta, and Carmona (1972) report the use of psychiatric patients as observers on a token economy ward. Point reinforcement was used to develop reliable observations of the behavior of fellow patients on the ward. Interestingly, the objective of training patients to collect data on others was to have their own behavior come under control of the behavior of the people they were observing. Aside from observing other patients, points were given for cooperative and critical verbal communication between observers vis à vis their recordings. The patient-assistants' tasks included not only the observation of others, but calculation and tabulation of the token earnings of others, and graph drawing. Patients have also been used in other fashions on the token economy ward. Heap et al. (1970) employed some patients who were in charge of supervising work assignments of others on the ward. Pomerleau et al. (1972) used patients to form a police force to maintain nonviolence on the ward. These patients were reinforced for serving in this capacity.

A more frequent use of patients on the ward is in developing group or peer social structure which supports and actively contributes to the appropriate behavior of individual patients. For example, in one psychiatric hospital (Pomerleau et al., 1972), patients were grouped into dyads. The dyads formed larger groups. Consequences in the program (e.g., fines) ordinarily imposed upon a single individual were shared by other members of the dyad or by other dyads. By making patients accountable to others for their behavior and by making the individual accountable to the group, it is hoped that the group will exert influence on behavior. Such an influence may well be adaptive once the program is withdrawn. Olson and Greenberg (1972) also have employed sharing of consequences with peers on a psychiatric ward. To exert group pressure, patients were allowed access to their funds (or coupons negotiable at the local canteen) on the basis of their own performance and the performance of the group to which they belonged. Performance of the group was evaluated on attendance to scheduled activities and weekly progress. The contribution of the group contingency to overall program efficacy was not determined.

Contingencies in which the performance of individuals results in group consequences have been carefully evaluated in nonpsychiatric settings. Wolf, Hanley, King, Lachowicz, and Giles (1970) used token reinforcement to develop attentive behavior in a classroom. For one child, out-of-seat behavior was not well controlled by individual reinforcement. A group contingency was devised so that tokens earned (i.e., not lost for inappropriate behavior) were divided among peers. Behavior improved markedly only when the reinforcers were shared.

Medland and Stachnik (1972) used a group contingency to control disruptive responses in a classroom. The class was divided into two teams. In a reversal design, phases were included in which groups received a point when a classroom rule was violated. Either or both team(s) earned extra minutes of recess if there were fewer than a certain number of points earned. This contingency was highly effective in reducing disruptive behavior. Walker and Buckley (1972) also showed that having peers share in token earnings or losses was effective in altering the behavior of individuals. In contingencies where there are group consequences for performance of the individual client, the precise influence of peer reinforcement and support is rarely evaluated.

Peers have been used in training pre-delinquent boys residing at Achievement Place, a home-style cottage in Kansas. In one report (Fixsen, Phillips, & Wolf, 1972) peer observations and self-observations were used to gather data on room-cleaning behaviors. Room-cleaning consisted of 21 specific behaviors which had to be recorded. Token reinforcement was used to increase reliability of agreement of observations between self and peer records. A dramatic utilization of peers in a token economy was reported by Bailey, Timbers, Phillips, and Wolf (1971). Some pre-delinquent boys served as therapists for others who had speech problems (articulation errors). Peers earned points for identifying words said correctly or incorrectly by the subjects. The subject either earned or lost points on the basis of his own performance. The peers effectively trained correct word pronunciation. These effects were obtained, although the peers had no specific speech training and almost no adult supervision. Only brief instructions were given, and adults remained absent from the sessions. The efficacy of the procedures was robust. In fact, the effects of speech training generalized to words not included in training and were maintained after two months.

The use of peers in reinforcement programs offers several potential advantages; such as, providing less restricted stimulus control over patient performance (which usually is consequated by the staff alone), providing more frequent reinforcement (planned and unplanned) for performance of target behaviors, freeing staff for supervisory functions thus better using their talents, and others (Kazdin, 1972a). Of course, the precise effects of using patients or clients in staff functions remain to be subjected to empirical scrutiny. An implicit hope may be that clients who perform some role in the programs of others may themselves change significantly (Dominguez et al., 1972). Presently, there is little empirical support justifying this strategy. However, since other advantages accrue to the use of peers, the effect on performance of those patients who are serving in staff functions is not the only consideration.

Use and Training of Staff

A major organizational change is evident in the role of staff (or nonprofessionals) in contact with the clients in a token economy. In operant programs, custodial care (Ullmann, 1967; Goffman, 1961) has been replaced with behavioral engineering (Ayllon & Michael, 1959) and contingency contracting (Homme, Csanyi, Gonzales, & Rechs, 1969). An increasing emphasis in token economies is being placed upon the behavior of staff who administer the program. Several authors have commented upon the performance of effective staff in administering operant programs (Gardner, 1973; Kazdin & Bootzin, 1972; Kazdin, 1973a; Liberman, 1968; O'Leary & Drabman, 1971; Turton & Gathercole, 1972). However, only quite recently have there been concerted efforts directed toward effecting behavior change in staff.

Staff training can be profitably viewed as a behavior modification program: target behaviors need to be identified; data need to be gathered to determine whether training has altered the target behaviors; the relationship between training and behavior change needs to be determined; and follow-up information needs to be gathered to insure that the gains are maintained. These conditions rarely obtain. Instructional or didactic procedures which are usually employed tend to have short-lived effects on the behavior of staff (Ayllon & Azrin, 1964; Azrin, Holz, Ulrich, & Goldiamond, 1961; Suchotliff et al., 1970), just as they do on the patients. However, recent creative efforts have been employed to alter staff behavior. Several behavioral techniques have been applied to this end, including reinforcement (Ayllon & Azrin, 1968; Wolf et al., 1968), reinforcement combined with punishment (McNamara, 1971), modeling (Cotler et al., 1972), feedback (Panyan, Boozer, & Morris, 1970), behavioral rehearsal (Gardner, 1970a), and self-paced instruction (Watson, Gardner, & Sanders, 1972). (See Gardner, 1973 for a review of staff training.)

A variety of attempts have been made to identify and apply reinforcers to consequate staff performance. In Ayllon and Azrin's (1965) token economy, contingent reinforcement (vacations, workshift preferences, bonuses) were delivered to staff for performance. On occasion, programs have used monetary reinforcers. For example, Pomerleau et al. (1972) employed a staff incentive system whereby the staff whose patients improved were reinforced with money. Data were gathered two days per week by the psychology staff to determine patient improvement and attendent consequences.

The use of strong reinforcers with staff would seem to be advisable. Katz, Johnson, and Gelfand (1972) evaluated the effects of instruction, verbal prompts, and monetary reinforcement in altering the behavior of psychiatric aides who administered a token economy. The target behaviors included administering social or tangible reinforcers to patients. Across five experimental phases, the results demonstrated that instructions, and prompts (reminders to staff to administer reinforcers) did not alter behavior. Staff behavior changed only when the monetary reinforcer ($15.00 bonus) was delivered for reaching a minimum criterion level.

This study carefully demonstrates a concern evinced in many programs. Staff behavior needs to be consequated. Bricker, Morgan, and Grabowski (1972) evaluated the effect of reinforcement and video feedback on the behavior of attendants working with retarded children. Staff received trading stamps contingent upon interacting with residents. Video taped records of staff behavior were shown to attendants. Reinforcers were delivered on the basis of quality and quantity of interactions with residents on the previous day. While the independent contribution of feedback, praise, knowledge of results and trading stamps could not be determined, interaction between attendants and residents improved dramatically through the use of this staff training regime.

Identifying reinforcers to be used with staff is no easy task. Most institutions do not allow wage bonuses, vacations, and promotions to be administered on the basis of staff performance on a token economy ward. Exigencies of the institution interfere with administration of such reinforcers. Praising staff and providing feedback for performance are possible alternatives, as feedback has been shown to maintain staff performance at high levels (Panyan et al., 1970). There is some concern, however, that administering intangible reinforcers is not easily monitored (Ayllon & Azrin, 1968). If praise or feedback are used, data need to be gathered to ensure that they are delivered systematically. Patterson, Cooke, and Liberman (1972) used a feedback newsletter to inform staff of programs on the ward and progress made in individual cases, and to point out staff achievements. By using regular newsletters, it's possible to guarantee feedback at least periodically.

Long-term maintenance of staff behavior remains an important goal. Even if staff training is effective initially, some means need to be developed to ensure that behavior is maintained. Once the training contingencies are withdrawn, staff performance returns to baseline or near baseline levels (Ackerman, 1972; Brown, Montgomery, & Barclay, 1969; Cooper, Thomson, & Baer, 1970; Kazdin, 1973a; Panyan et al., 1970). The issue of long-term changes in staff behavior has received little investigation.

Investigators have begun to look at what effects token programs for the clients have on the behavior of the staff. McReynolds and Coleman (1972) evaluated attitude changes made in staff working on a token ward for psychiatric patients. After one year of the program, staff opinion increased in the number of patients that were considered responsive to staff members and capable of extra-hospital adjustment. In a comparison, attitudes of staff on the token ward were more favorable regarding the potential benefits of treatment of patients, than were the attitudes of staff on a similar ward. Further, staff on the token ward rated a larger percentage of patients as aware of themselves and their behavior, and as experiencing normal human emotions than did staff on the comparison ward.

In other settings, the effect of a token program for clients has had desirable effects on staff behavior. Chadwick and Day (1971) found that having teacher and classroom aides administer tokens for academic behavior increased their use of approval, and decreased their use of disapproval during the phases in which tokens were administered. Breyer and Allen (1972) also demonstrated that implementing a token program in a classroom altered the teacher's use of praise and reprimands. Other procedures employed to alter the teacher's behavior (e.g., training and feedback) had not been very effective.

A major feature of token programs is that they structure staff behavior. Not only are certain behaviors more likely to be emitted but delivery of contingent token reinforcement provides an increase in the rate of social interaction with the clients (Mandelker, Brigham, & Bushell, 1970). As Ferster (1972) suggests, token programs may be effective because of the way in which staff are more observant of, and sensitive to, the clients' behavior, including client responses not directly included in the token reinforcement contingencies. Certainly a closer look at staff performance needs to be made.

Role of the Community

The delivery of mental health services through operant procedures provides a new role for the community. Treatment needs to be enmeshed directly into community life so that there is a systematic transition from treatment to extra-hospital existence (Fairweather, Sanders, Maynard, Cressler & Bleck, 1969). Relatively few token programs have included "community adjustment" into the contingencies.

Henderson (1969, 1970) described a facility for psychotic men situated in the community. Tokens were delivered for social, vocational, and counter-symptom behaviors. Various aspects of community life were incorporated into the treatment of patients. For example, reinforcement was delivered for social interaction, including interacting with visitors (volunteers) from the community (Henderson & Scoles, 1970). Several levels of social involvement were differentially reinforced, including; superficial participation, interpersonal transactions indicative of social involvement, and initiation of conversation. Social skills were reinforced during activities in the facility and in the community (e.g., at YMCA). In addition, activities of patients such as securing job interviews or employment were reinforced. Employers were involved in the program by providing feedback so that consequences would be given for community performance (Kelley & Henderson, 1971).

Community involvement has included a host of paraprofessionals who contribute greatly to the habilitation of individuals who are housed in various treatment facilities (Ayllon & Wright, 1972). Bailey, Wolf, and Phillips (1970) provided token reinforcement, on the basis of academic performance at school, for pre-delinquent boys living in a home-style cottage. The teacher made daily reports on cards which determined privileges earned at the home. The success of the program depended upon the classroom teacher's accurately reporting behavior so that consequences could be applied contingently in the training facility. The teacher also participated in shifting subjects from continuous to intermittent reinforcement by gradually fading the use of the report card.

In a school setting, McKenzie, Clark, Wolf, Kothera, & Benson (1968) reinforced academic behaviors in a class of children with learning disabilities. Weekly grades were given as reinforcers. Parents were trained to systematically praise good grades and mildly disapprove of poor grades. Eventually parents gave allowances contingent upon grade performance. Classroom academic performance appeared to change during the program. An interesting feature, of course, is that parent participation contributed substantially.

Token programs are increasingly more frequent in the home. Therapists are consulting with parents on an outpatient basis to develop token programs for children. Christopherson, Arnold, Hill, and Quilitch (1972) instructed two sets of parents to implement token programs to alter behavior of their children. In one family, three children were reinforced for making their beds, going to bed on time, and not whining or bickering. Using a multiple-baseline design across behaviors, the effects of token reinforcement and response cost were carefully demonstrated. Other investigators have shown dramatic effects in training parents as experimenters and observers in altering child behaviors (Hall, Axelrod, Tyler, Grief, Jones, & Robertson, 1972). In many cases relatively little consulting time is required. A few hours in explaining procedures can be supplemented by phone calls. Some parents can be trained in a matter of a few sessions (Christopherson et al., 1972).

Increasingly, relatives, employers, teachers, and peers, are called upon to alter behavior of subjects whose behavior has been labeled, or singled out, as deviant. The use of treatment programs can often be carried out in the home to avoid institutionalization. After institutionalization has taken place, or the client is participating in some program (e.g., educational), the efficacy of the procedures can be enhanced by continuing the use of the contingencies in the community and home. The implications for the use of nonprofessionals include preventive treatment of individuals who are prospective patients, as well as follow-up on patients who are discharged. An increased emphasis on prevention and follow-up can drastically alter the necessity for mental health services.

The implications of token economy research extend beyond community involvement in the rehabilitation or treatment of an individual whose behavior has been labeled abnormal. Token programs represent a microcosm where economic features, issues of government, law, and participation of the governed, all become salient. Research resulting from token economies may offer a great deal of information about social organization and individual behavior. Skinner (1948) designed an elaborate token economy which extends far beyond the current systematic use of operant principles. Of course, at present the extent to which social change may result from work in miniature societies is unclear. The intermediate goals of widespread reduction of deviant behavior and long-term therapeutic change have not yet been attained. However, the current trends in token economies suggest major departures in the control of deviant behavior and the role accorded society in achieving control.

REFERENCES

Ackerman, J. M. *Operant Conditioning Techniques for the Classroom Teacher.* Glenview, Illinois: Scott, Foresman and Company, 1972.

Allen, D. J., & Magaro, P. A. "Measures of change in token-economy programs." *Behaviour Research and Therapy*, 1971, 9, 311-318.

Arann, L., & Horner, V. M. "Contingency management in an open psychiatric ward." *Journal of Behavior Therapy and Experimental Psychiatry*, 1972, 3, 31-37.

Atthowe, J. M., & Krasner, L. "Preliminary report on the application of contingent reinforcement procedures (token economy) on a 'chronic' psychiatric ward." *Journal of Abnormal Psychology*, 1968, 73, 37-43.

Ayllon, T. "Intensive treatment of psychotic behavior by stimulus satiation and food reinforcement." *Behaviour Research and Therapy*, 1963, 1, 53-62.

Ayllon, T. "Some behavioral problems associated with eating in chronic schizophrenic patients." In L. P. Ullmann & L. Krasner (Eds.), *Case Studies in Behavior Modification.* New York: Holt, Rinehart, & Winston, 1965. Pp. 73-77.

Ayllon, T., & Azrin, N. H. "Reinforcement and instructions with mental patients." *Journal of the Experimental Analysis of Behavior*, 1964, 7, 327-331.

Ayllon, T., & Azrin, N. H. "The measurement and reinforcement of behavior of psychotics." *Journal of the Experimental Analysis of Behavior*, 1965, 8, 357-383.

Ayllon, T., & Azrin, N. H. *The Token Economy: A Motivational System for Therapy and Rehabilitation.* New York: Appleton-Century-Crofts, 1968.

Ayllon, T., & Haughton, E. "Control of the behavior of schizophrenic patients by food." *Journal of the Experimental Analysis of Behavior*, 1962, 5, 343-352.

Ayllon, T., & Haughton, E. "Modification of symptomatic verbal behavior of mental patients." *Behaviour Research and Therapy*, 1964, 2, 87-97.

Ayllon, T., & Michael, J. "The psychiatric nurse as a behavioral engineer." *Journal of the Experimental Analysis of Behavior*, 1959, 3, 323-334.

Ayllon, T., & Wright, P. "New roles for the paraprofessional." In S. W. Bijou & E. Ribes-Inesta (Eds.), *Behavior Modification: Issues and Extensions.* New York: Academic Press, 1972, Pp. 115-125.

Azrin, N. H., & Holz, W. C. "Punishment." In W. K. Honig (Ed.), *Operant Behavior: Areas of Research and Application.* New York: Appleton-Century-Crofts, 1966. Pp. 380-447.

Azrin, N. H., Holz, W., Ulrich, R., & Goldiamond, I. "The control of the content of conversation through reinforcement." *Journal of the Experimental Analysis of Behavior*, 1961, 4, 25-30.

Azrin, N. H., & Lindsley, O. R. "The reinforcement of cooperation between children." *Journal of Abnormal and Social Psychology*, 1956, 52, 100-102.

Baer, D. M., Wolf, M. M., & Risley, T. R. "Some current dimensions of applied behavior analysis." *Journal of Applied Behavior Analysis*, 1968, 1, 91-97.

Bailey, J. S., Timbers, G. D., Phillips, E. L., & Wolf, M. M. "Modification of articulation errors of pre-delinquents by their peers." *Journal of Applied Behavior Analysis*, 1971, 4, 265-281.

Bailey, J. S., Wolf, M. M., & Phillips, E. L. "Home-based reinforcement and the modification of pre-delinquents' classroom behavior." *Journal of Applied Behavior Analysis*, 1970, 3, 223-233.

Bandura, A., Blanchard, E. B., & Ritter, B. "Relative efficacy of desensitization and modeling approaches for inducing behavioral, affective, and attitudinal changes." *Journal of Personality and Social Psychology*, 1969, 13, 173-199.

Barrett, B. H., & Lindsley, O. R. "Deficits in acquisition of operant discrimination and differentiation shown by institutionalized retarded children." *American Journal of Mental Deficiency*, 1962, 67, 424-436.

Bartlett, D., Ora, J. P., Brown, E., & Butler, J. "The effects of reinforcement on psychotic speech in a case of early infantile autism, age 12." *Journal of Behavior Therapy and Experimental Psychiatry*, 1971, 2, 145-149.

Bennett, P. S., & Maley, R. F. "Modification of interactive behaviors in chronic mental patients." *Journal of Applied Behavior Analysis*, 1973, in press.

Bijou, S. W., Peterson, R. F., Harris, F. R., Allen, K. E., & Johnston, M. S. "Methodology for experimental studies of young children in natural settings." *Psychological Record*, 1969, 19, 177-210.

Birky, H. J., Cambliss, J. E., & Wasden, R. "A comparison of residents discharged from a token economy and two traditional psychiatric programs." *Behavior Therapy*, 1971, 2, 46-51.

Bolstad, O. D., & Johnson, S. M. "Self-regulation in the modification of disruptive behavior." *Journal of Applied Behavior Analysis*, 1972, 5, 443-454.

Boren, J. J., & Colman, A. D. "Some experiments on reinforcement principles within a psychiatric ward for delinquent soldiers." *Journal of Applied Behavior Analysis*, 1970, 3, 29-37.

Breyer, N. L., & Allen, G. J. "Effects of implementing a token economy on teacher attending behavior." Paper presented at Sixth Annual Meeting of the Association for the Advancement of Behavior Therapy, New York, October, 1972.

Bricker, W. A., Morgan, D. G., & Grabowski, J. G. "Development and maintenance of a behavior modification repertoire of cottage attendants through TV feedback." *American Journal of Mental Deficiency*, 1972, 77, 128-136.

Brown, J., Montgomery, R., & Barclay, J. "An example of psychologist management of teacher reinforcement procedures in the elementary classroom." *Psychology in the Schools*, 1969, 6, 336-340.

Burchard, J. D., & Barrera, F. "An analysis of time-out and response cost in a programmed environment." *Journal of Applied Behavior Analysis*, 1972, 5, 271-282.

Burchard, J. D., & Tyler, V. O. "The modification of delinquent behaviour through operant conditioning." *Behaviour Research and Therapy*, 1965, 2, 245-250.

Carlson, C. G., Hersen, M., & Eisler, R. M. "Token economy programs in the treatment of hospitalized adult psychiatric patients." *Journal of Nervous and Mental Disease*, 1972, 155, 192-204.

Carpenter, P., & Caron, R. "Green stamp therapy: Modification of delinquent behavior through food trading stamps." *Proceedings, 76th Annual Convention, American Psychological Association*, 1968, 3, 531-532.

Chadwick, B. A., & Day, R. C. "Systematic reinforcement: Academic performance of underachieving students." *Journal of Applied Behavior Analysis*, 1971, 4, 311-319.

Christopherson, E. R., Arnold, C. M., Hill, D. W., & Quilitch, H. R. "The home point system: Token reinforcement procedures for application by parents of children with behavior problems." *Journal of Applied Behavior Analysis*, 1972, 5, 485-497.

Cohen, R., Florin, I., Grusche, A., Meyer-Osterkamp, S., & Sell, H. "The introduction of a token economy in a psychiatric ward with extremely withdrawn chronic schizophrenics." *Behaviour Research and Therapy*, 1972, 10, 69-74.

Cooper, M. L., Thomson, C. L., & Baer, D. M. "The experimental modification of teacher attending behavior." *Journal of Applied Behavior Analysis*, 1970, 3, 153-157.

Cotler, S. B., Applegate, G., King, L. W., Kristal, S. "Establishing a token economy program in a state hospital classroom: A lesson in training student and teacher." *Behavior Therapy*, 1972, 3, 209-222.

Davison, G. C. "Appraisal of behavior modification techniques with adults in institutional settings." In C. M. Franks (Ed.), *Behavior Therapy: Appraisal and Status*. New York: McGraw-Hill, 1969. Pp. 220-278.

Dominguez, B., Acosta, T. F., & Carmona, D. "Discussion: A new perspective: Chronic patients as assistants in a behavior rehabilitation program in a psychiatric institution." In S. W. Bijou & E. Ribes-Inesta (Eds.), *Behavior Modification: Issues and Extensions*. New York: Academic Press, 1972. Pp. 127-132.

Ellsworth, J. R. "Reinforcement therapy with chronic patients." *Hospital and Community Psychiatry*, 1969, 20, 36-38.

Eysenck, H. J. (Ed.) *Experiments in Behavior Therapy*. London: Pergamon Press, 1964.

Fairweather, G. W., Sanders, D. H., Maynard, H., Cressler, D. L., & Bleck, D. S. *Community Life for the Mentally Ill: An Alternative to Institutional Care*. Chicago: Aldine, 1969.

Ferritor, D. E., Buckholdt, D., Hamblin, R. L., & Smith, L. "The noneffects of contingent reinforcement for attending behavior on work accomplished." *Journal of Applied Behavior Analysis,* 1972, 5, 7-17.

Ferster, C. B. "An experimental analysis of clinical phenomena." In S. W. Bijou & E. Ribes-Inesta (Eds.), *Behavior Modification: Issues and Extensions.* New York: Academic Press, 1972, Pp. 134-147.

Ferster, C. B., & DeMyer, M. K. "The development of performances in autistic children in an automatically controlled environment." *Journal of Chronic Diseases,* 1961, 13, 312-345.

Ferster, C. B., & DeMyer, M. K. "A method for the experimental analysis of the behavior of autistic children." *American Journal of Orthopsychiatry,* 1962, 1, 87-110.

Fetke, G. C. "The relevance of economic theory and technology to token reinforcement systems: A comment." *Behaviour Research and Therapy,* 1972, 10, 191-192.

Fixsen, D. L., Phillips, E. L., & Wolf, M. M. "Achievement Place: The reliability of self-reporting and peer-reporting and their effects on behavior." *Journal of Applied Behavior Analysis,* 1972, 5, 19-30.

Freeman, H. E., & Simmons, O. G. *The Mental Patient Comes Home.* New York: Wiley, 1963.

Gardner, J. M. "Teaching behavior modification to nonprofessionals." *Journal of Applied Behavior Analysis,* 1972, 5, 517-521.

Gardner, J. M. "Training the trainers: A review of research on teaching behavior modification." In C. M. Franks & R. Rubin (Eds.), *Progress in Behavior Therapy,* 1971.

Garlington, W. K., & Lloyd, K. E. "The establishment of a token economy ward at the State Hospital North in Orofino, Idaho." Paper presented at research meeting, Fort Steilacoom, Washington, November, 1966.

Glynn, E. L. "Classroom applications of self-determined reinforcement." *Journal of Applied Behavior Analysis,* 1970, 3, 123-132.

Goffman, E. *Asylums.* New York: Anchor, 1961.

Gripp, R. F., & Magaro, P. A. "A token economy program evaluation with untreated control ward comparisons." *Behaviour Research and Therapy,* 1971, 9, 137-149.

Hall, R. V., Axelrod, S., Tyler, L., Grief, E., Jones, F. C., & Robertson, R. "Modification of behavior problems in the home with a parent as observer and experimenter." *Journal of Applied Behavior Analysis,* 1972, 5, 53-64.

Hartlage, L. C. "Subprofessional therapists's use of reinforcement versus traditional psychotherapeutic techniques with schizophrenics." *Journal of Consulting and Clinical Psychology,* 1970, 34, 181-183.

Hauserman, N., Zweback, S., & Plotkin, A. "Use of concrete reinforcement to facilitate verbal initiations in adolescent group therapy." *Journal of Consulting and Clinical Psychology,* 1972, 38, 90-96.

Hawkins, R. P., Peterson, R. F., Schweid, E., & Bijou, S. W. "Behavior therapy in the home: Amelioration of problem parent-child relations with the parent in a therapeutic role." *Journal of Experimental Child Psychology,* 1966, 4, 99-107.

Heap, R. F., Boblitt, W. E., Moore, C. H., & Hord, J. E. "Behavior-milieu therapy with chronic neuropsychiatric patients." *Journal of Abnormal Psychology*, 1970, 76, 349-354.

Henderson, J. D. "The use of dual reinforcement in an intensive treatment system." In R. D. Rubin & C. M. Franks (Eds.), *Advances in Behavior Therapy*, 1968. New York: Academic Press, 1969.

Henderson, J. D. & Scoles, P. E. "A community based behavioral operant environment for psychotic men." *Behavior Therapy*, 1970, 1, 245-251.

Hewett, F. M., Taylor, F. D., & Artuso, A. A. "The Santa Monica Project: Evaulation of an engineered classroom design with emotionally disturbed children." *Exceptional Children*, 1969, 35, 523-529.

Homme, L., Csanyi, A., Gonzales, M., & Rechs, J. *How to Use Contingency Contracting in the Classroom.* Champaign, Illinois: Research Press, 1969.

Horton, L. E. "Generalization of aggressive behavior in adolescent delinquent boys." *Journal of Applied Behavior Analysis*, 1970, 3, 205-211.

Ingham, R., Andrews, G. J., & Winkler, R. C. "The application of token reinforcement systems to the control of stuttering." *Journal of Speech and Hearing Disorders*, 1973, in press.

Issacs, W., Thomas, J., & Goldiamond, I. "Application of operant conditioning to reinstate verbal behavior in psychotics." *Journal of Speech and Hearing Disorders*, 1960, 25, 8-12.

Johnson, S. M. "Effects of self-reinforcement and external reinforcement in behavior modification." *Proceedings, 77th Annual Convention, American Psychological Association*, 1969, 4, 535-536.

Johnson, S. M. "Self-reinforcement vs. external reinforcement in behavior modification with children." *Developmental Psychology*, 1970, 3, 147-148.

Johnston, J. M. "Punishment of human behavior." *American Psychologist*, 1972, 27, 1033-1054.

Kagel, J. H., & Winkler, R. C. "Behavioral economics: Areas of cooperative research between economies and applied behavior analysis." *Journal of Applied Behavior Analysis*, 1972, 5, 335-342.

Kanfer, F. H. "Self-regulation: Research, issues, and speculations." In C. Neuringer & J. L. Michael (Eds.), *Behavior Modification in Clinical Psychology*, New York: Appleton-Century-Crofts, 1970, Pp. 178-220.

Katz, R. C., Johnson, C. A., & Gelfand, S. "Modifying the dispensing of reinforcers: Some implications for behavior modification with hospitalized patients." *Behavior Therapy*, 1972, 3, 579-588.

Kaufman, K. F., & O'Leary, K. D. "Reward, cost, and self-evaluation procedures for disruptive adolescents in a psychiatric hospital school." *Journal of Applied Behavior Analysis*, 1972, 5, 293-309.

Kazdin, A. E. "The effect of response cost in suppressing behavior in a pre-psychotic retardate." *Journal of Behavior Therapy and Experimental Psychiatry*, 1971, 2, 137-140. (a)

Kazdin, A. E. "Toward a client administered token reinforcement program." *Education and Training of the Mentally Retarded*, 1971, 6, 52-55. (b)

Kazdin, A. E. "Implementing token programs: The use of staff, patients, and the institution of maximizing change." Paper presented at the Sixth Annual Meeting of the Association for the Advancement of Behavior Therapy, New York, October, 1972. (a)

Kazdin, A. E. "Response cost: The removal of conditioned reinforcers for therapeutic change." *Behavior Therapy*, 1972, 3, 533-546. (b)

Kazdin, A. E. "The assessment of teacher training in a reinforcement program." *Journal of Teacher Education*, 1973, in press. (a)

Kazdin, A. E. "The effect of response cost and aversive stimulation in suppressing punished and nonpunished speech disfluencies." *Behavior Therapy*, 1973, in press. (b)

Kazdin, A. E. "The effect of vicarious reinforcement on attentive behavior in the classroom." *Journal of Applied Behavior Analysis*, 1973, in press. (c)

Kazdin, A. E. "The failure of some patients to respond to token programs." *Journal of Behavior Therapy and Experimental Psychiatry*, 1973, in press. (d)

Kazdin, A. E. "Methodological and assessment considerations in evaluating reinforcement programs in applied settings." *Journal of Applied Behavior Analysis*, 1973, in press. (e)

Kazdin, A. E. "The role of instructions and reinforcement in behavior changes in token reinforcement programs." *Journal of Educational Psychology*, 1973, in press. (f)

Kazdin, A. E., & Bootzin, R. R. "The token economy: An evaluative review." *Journal of Applied Behavior Analysis*, 1972, 5, 343-372.

Kazdin, A. E. & Polster, R. "Intermittent token reinforcement and response maintenance in extinction." *Behavior Therapy*, 1973, in press.

Kelleher, R. T., & Gollub, L. R. "A review of positive conditioned reinforcement." *Journal of the Experimental Analysis of Behavior*, 1962, 5, 543-597.

Kelley, K. M., & Henderson, J. D. "A community-based operant learning environment II: Systems and procedures." In R. D. Rubin, H. Fensterheim, A. A. Lazarus, & C. M. Franks, (Eds.), *Advances in Behavior Therapy*, New York: Academic Press, 1971.

King, G. F., Armitage, S. G., & Tilton, J. R. "A therapeutic approach to schizophrenics of extreme pathology: An operant-interpersonal method." *Journal of Abnormal and Social Psychology*, 1960, 61, 276-286.

Kubany, E. S., Weiss, L. E., & Sloggett, B. B. "The good behavior clock: A reinforcement/time-out procedure for reducing disruptive classroom behavior." *Journal of Behavior Therapy and Experimental Psychiatry*, 1971, 2, 173-179.

Leitenberg, H. "Is time-out from positive reinforcement an aversive event? A review of the experimental evidence." *Psychological Bulletin*, 1965, 64, 428-441.

Leitenberg, H., Wincze, J., Butz, R., Callahan, E., & Agras, W. "Comparison of the effect of instructions and reinforcement in the treatment of a neurotic avoidance response: A single case experiment." *Journal of Behavior Therapy and Experimental Psychiatry*, 1970, 1, 53-58.

Liberman, R. "A view of behavior modification projects in California." *Behaviour Research and Therapy*, 1968, 6, 331-341.

Liberman, R. P. "Reinforcement of social interaction in a group of chronic mental patients." In R. D. Rubin, H. Fensterheim, J. D. Henderson, & L. P. Ullmann (Eds.),

Advances in Behavior Therapy. New York: Academic Press, 1972. Pp. 151-159.

Lindsley, O. R. "Characteristics of the behavior of chronic psychotics as revealed by free-operant conditioning methods." *Diseases of the Nervous System,* (Monograph Supplement), 1960, 31, 66-78.

Lloyd, K. E., & Abel, L. "Performance on a token economy psychiatric ward: A two-year summary." *Behaviour Research and Therapy,* 1970, 8, 1-9.

Lloyd, K. E., & Garlington, W. K. "Weekly variations in performance on a token economy psychiatric ward." *Behaviour Research and Therapy,* 1968, 6, 407-410.

Lovitt, T. C., & Curtiss, K. A. "Academic response rate as a function of teacher- and self-imposed contingencies." *Journal of Applied Behavior Analysis,* 1969, 2, 49-53.

Maley, R. F., Feldman, G. L., & Ruskin, R. S. "Evaluation of patient improvement in a token economy treatment program." *Journal of Abnormal Psychology,* 1973, 82, 141-144.

Mandelker, A. V., Brigham, T. A., & Bushell, D. "The effects of token procedures on a teacher's social contracts with her students." *Journal of Applied Behavior Analysis,* 1970, 3, 169-174.

Marks, J., Sonoda, B., & Schalock, R. "Reinforcement vs. relationship therapy for schizophrenics." *Journal of Abnormal Psychology,* 1968, 73, 379-402.

McKenzie, H. S., Clark, M., Wolf, M. M., Kothera, R., & Benson, C. "Behavior modification of children with learning disabilities using grades as tokens and allowances as back-up reinforcers." *Exceptional Children,* 1968, 745-752.

McNamara, H. J., & Wike, E. L. "The effects of irregular learning conditions upon the rate and permanence of learning." *Journal of Comparative and Physiological Psychology,* 1958, 51, 363-366.

McNamara, J. R. "Teacher and students as a source for behavior modification in the classroom." *Behavior Therapy,* 1971, 2, 205-213.

McReynolds, W. T., & Coleman, J. "Token economy: Patient and staff changes." *Behaviour Research and Therapy,* 1972, 10, 29-34.

Medland, M. B., & Stachnik, T. J. "Good-behavior game: A replication and systematic analysis." *Journal of Applied Behavior Analysis,* 1972, 5, 45-51.

Meichenbaum, D. H. "The effects of instruction and reinforcement on thinking and language behaviour of schizophrenics." *Behaviour Research and Therapy,* 1969, 7, 101-114.

Meichenbaum, D., & Cameron, R. "Training schizophrenics to talk to themselves: A means of developing attentional controls." *Behavior Therapy,* in press.

Meichenbaum, D., & Goodman, J. "Training impulsive children to talk to themselves: A means of developing self-control." *Journal of Abnormal Psychology,* 1971, 77, 115-126.

Milby, J. B. "A brief review of token economy treatment." Paper presented at 70th Annual Psychological Association, Honolulu, Hawaii, September, 1972.

Mulligan, W., Kaplan, R. D., & Reppucci, N. D. "Changes in cognitive variables among behavior problem elementary school boys treated in a token economy special classroom." Paper presented at Fifth Annual Meeting of the Association for the Advancement of Behavior Therapy, Washington, D.C., September, 1971.

O'Brien, F., & Azrin, N. H. "Symptom reduction by functional displacement in a token economy: A case study." *Journal of Behavior Therapy and Experimental Psychiatry*, 1972, 3, 205-207.

O'Leary, K. D., & Drabman, R. "Token reinforcement programs in the classroom: A review." *Psychological Bulletin*, 1971, 75, 379-398.

Olson, R. P., & Greenberg, D. J. "Effects of contingency contracting and decision-making groups with chronic mental patients." *Journal of Consulting and Clinical Psychology*, 1972, 38- 376-383.

Panyan, M., Boozer, H., & Morris, N. "Feedback to attendants as a reinforcer for applying operant techniques." *Journal of Applied Behavior Analysis*, 1970, 3, 1-4.

Parrino, J. J., George, L., & Daniels, A. C. "Token control of pill-taking behavior in a psychiatric ward." *Journal of Behavior Therapy and Experimental Psychiatry*, 1971, 2, 181-185.

Patterson, R., Cooke, C., & Liberman, R. P. "Reinforcing the reinforcers: A method of supplying feedback to nursing personnel." *Behavior Therapy*, 1972, 3, 444-446.

Paul, G. L. "Insight versus desensitization in psychotherapy two years after termination." *Journal of Consulting Psychology*, 1967, 31, 333-348.

Paul, G. L. "Chronic mental patient: Current status — future directions." *Psychological Bulletin*, 1969, 71, 81-94.

Phillips, E. L., Phillips, E. A., Fixsen, D. F., & Wolf, M. M. "Achievement Place: Modification of the behaviors of pre-delinquent boys within a token economy." *Journal of Applied Behavior Analysis*, 1971, 4, 45-59.

Pomerleau, O. F., Bobrove, P. H., & Harris, L. C. "Some observations on a controlled social environment for psychiatric patients." *Journal of Behavior Therapy and Experimental Psychiatry*, 1972, 3, 15-21.

Reisinger, J. J. "The treatment of 'anxiety-depression' via positive reinforcement and response cost." *Journal of Applied Behavior Analysis*, 1972, 5, 125-130.

Schaefer, H. H., & Martin, P. L. "Behavioral therapy for 'apathy' of hospitalized schizophrenics." *Psychological Reports*, 1966, 19, 1147-1158.

Schaefer, H. H., & Martin, P. L. *Behavioral Therapy*. New York: McGraw-Hill, 1969.

Shean, J. D., & Zeidberg, Z. "Token reinforcement therapy: A comparison of matched groups." *Journal of Behavior Therapy and Experimental Psychiatry*, 1971, 2, 95-105.

Sherman, J. A., & Baer, D. M. "Appraisal of operant therapy techniques with children and adults." In C. M. Franks (Ed.), *Behavior Therapy: Appraisal and Status*. New York: McGraw-Hill, 1969. Pp. 192-219.

Skinner, B. F. *Walden Two*. New York: Macmillan, 1948.

Staats, A. W. "A general apparatus for the investigation of complex learning in children." *Behaviour Research and Therapy*, 1968, 6, 45-50.

Staats, A. W., & Butterfield, W. H. "Treatment of nonreading in a culturally deprived juvenile delinquent: An application of learning principles." *Child Development*, 1965, 4, 925-942.

Staats, A. W., Minke, K. A., Finley, F. R., Wolf, M., & Brooks, L. O. "A reinforcer system and experimental procedure for the laboratory study of reading acquisition." *Child Development*, 1964, 35, 209-231.

Staats, A. W., Staats, C. K., Schultz, R. E., & Wolf, M. "The conditioning of textual responses using 'extrinsic' reinforcers." *Journal of the Experimental Analysis of Behavior*, 1962, 5, 33-40.

Stayer, S. J., & Jones, F. "Ward 108: Behavior modification and the delinquent soldier." Paper presented at Behavioral Engineering Conference, Walter Reed General Hospital, 1969.

Steffy, R. A., Hart, J., Craw, M., Torney, D., & Marlett, N. "Operant behaviour modification techniques applied to severely regressed and aggressive patients." *Canadian Psychiatric Association Journal*, 1969, 14, 59-67.

Striefel, S. "Time-out and concurrent fixed-ratio schedules with human subjects." *Journal of the Experimental Analysis of Behavior*, 1972, 17, 213-219.

Suchotliff, L., Greaves, S., Stecker, H., & Berke, R. "Critical variables in the token economy." *Proceedings of the 78th Annual Convention of the American Psychological Association*, 1970, 5, 517-518.

Surratt, P. R., Ulrich, R. E., & Hawkins, R. P. "An elementary student as a behavioral engineer." *Journal of Applied Behavior Analysis*, 1969, 2, 85-92.

Turton, B. K., & Gathercole, C. E. "Token economies in the U. K. and Eire." *Bulletin of the British Psychological Society*, 1972, 25, 83-87.

Twardosz, S., & Sajwaj, T. "Multiple effects of a procedure to increase sitting in a hyperactive retarded boy." *Journal of Applied Behavior Analysis*, 1972, 5, 73-78.

Ullmann, L. P. *Institution and Outcome: A Comparative Study of Psychiatric Hospitals.* London: Pergamon Press, 1967.

Ullmann, L. P., & Krasner, L. (Eds.) *Case Studies in Behavior Modification.* New York: Holt, Rinehart, & Winston, 1965.

Upper, D. "A 'ticket' system for reducing ward rules violations on a token economy program." Paper presented at Association for the Advancement of Behavior Therapy, Washington, D.C., September, 1971.

Wahler, R. G. "Behavior therapy for oppositional children: Love is not enough." Paper read at Eastern Psychological Association Meeting, Washington, D. C., April, 1968.

Walker, H. M., & Buckley, N. K. "Programming generalization and maintenance of treatment effects across time and across settings." *Journal of Applied Behavior Analysis*, 1972, 5, 209-224.

Walker, H. M., Mattson, R. H., & Buckley, N. K. "The functional analysis of behavior within an experimental class setting." In W. C. Becker (Ed.), *An Empirical Basis for Change in Education.* Chicago: Science Research Associates, 1971. Pp. 236-263.

Watson, L. S., Gardner, J. M., & Sanders, C. "Shaping and maintaining behavior modification skills in staff members in an MR institution: Columbus State Institute Behavior Modification Program." *Mental Retardation*, 1971, 9, 39-42.

Whitman, T. L., Mercurio, J. R., & Caponigri, V. "Development of social responses in two severely retarded children." *Journal of Applied Behavior Analysis*, 1970, 3, 133-138.

Wickes, I. G. "Treatment of persistent enuresis with the electric buzzer." *Archives of Disease in Childhood*, 1958, 33, 160-164.

Wincze, J. P., Leitenberg, H., & Agras, W. S. "The effects of token reinforcement and feedback on the delusional verbal behavior of chronic paranoid schizophrenics." *Journal of Applied Behavior Analysis*, 1972, 5, 247-262.

Winkler, R. C. "Management of chronic psychiatric patients by a token reinforcement system." *Journal of Applied Behavior Analysis*, 1970, 3, 47-55.

Winkler, R. C. "Reinforcement schedules for individual patients in a token economy." *Behavior Therapy*, 1971, 2, 534-537.

Winkler, R. C. "A theory of equilibrium in token economies." *Journal of Abnormal Psychology*, 1972, 79, 169-173.

Winkler, R. C., & Krasner, L. "The contribution of economics to token economies." Paper presented at meeting of Eastern Psychological Association, New York, April, 1971.

Wolf, M. M., Giles, D. K., & Hall, R. V. "Experiments with token reinforcement in a remedial classroom." *Behaviour Research and Therapy*, 1968, 6, 51-64.

Wolf, M. M., Hanley, E. L., King, L. A., Lachowicz, J., & Giles, D. K. "The timer game: A variable interval contingency for the management of out-of-seat behavior." *Exceptional Children*, 1970, 37, 113-117.

SECTION IV

Accountability and Systems of Behavior

Two recent trends in the history of mental health have led to major and far-reaching changes in the structure of mental health delivery systems. First, the abundance and relative ease and availability of resources for human services in general, and mental health services in particular, have been sharply reduced and constrained. Instead of conceptualizing program development within a framework of an expanding resource base, planners and administrators have increasingly been required to think within the constraints of stabilized, sometimes decreasing, levels of resources. This has led to both an increased level of competition between agencies for available resources, and the development of superordinate systems designed to make as rational as possible the allocation of resources among human service organizations.

A second recent trend has been the rapid expansion of behaviorally based strategies of mental health treatment, and as a corollary to this trend, the development of behaviorally defined outcome measures for mental health treatment programs. To be sure the latter is still relatively primitive. However, the mystique has been removed from "the fifty-minute hour". The demand for clear and specifiable treatment outcomes, and the linking of those outcomes to the systematic use of resources for treatment purposes is growing rapidly.

Taken together, these trends have made it possible for the problem of accountability in mental health programs to become a legitimate and viable concern. That is, the political problems of resource management, coupled with the development of treatment technologies that are increasingly overt, specifiable, and operational, have created certain social conditions. Within these, searching questions regarding the benefits or service value of resources expended can legitimately be asked, and the answers expected. Even as late as 1967 or 1968 this was not generally the case.

A host of conceptual, technological, methodological and political problems will have to be solved if effective systems of accountability are to be developed for mental health treatment systems. However, it is not as if these systems have in the past been without systems of accountability, as researchers, administrators, and planners often mistakenly assume. What is presently before us is the restructuring *of existing accountability practices and procedures; a movement from disjointed, often irrelevant, outcome measures to more rational, adaptive, outcome orientations. Various sorts of accountability practices have been with us for years. For example, legislators have held mental hospitals and state departments of mental health responsible for keeping "undesirable elements" off the streets of their communities, and for providing patronage jobs. Similarly, community mental health centers have been held responsible by their boards for client volume rather than for treatment modalities leading to particular kinds of treatment outcomes.*

What we are suggesting here is that attempts to generate methods and practices leading to more rational methods of joining resource allocation and outcome measures will be seen as involving major changes in the exercise of power. Legislators, for example, may view such a move as power being transferred to those who read cost-benefit print-outs from computer programs. Agency administrators may view operational accountability practices as increasing the power of a few specialized professionals; an organizational change which may be seen as limiting their decision flexibility.

We would propose that changes in accountability practices, moving from informal to formal rules, and from unspecified to specified procedures and outcomes, are forms of organizational change that will render it possible for persons in decision making positions to more rationally make program decisions. Further, it is likely that this rationality will lead both to more efficient resource utilization and to more effective services. The primary beneficiary will be the consumer of services.

Three major problems have been identified in the development of management information systems designed to increase accountability. The first of these has to do with what is actually measured as an outcome of a service rendered. The problem, of course, is that interesting but irrelevant behaviors are likely to be carefully counted. In the past, there has been a strong tendency for program evaluation efforts to focus on visible, tangible, and measurable processes, but for those measures to be largely unrelated to the actual outcomes of services. For example, management information systems in mental health have typically focused on such measures as the number of clients seen per agency per month, diagnostic category of clients, number of hours of client treatment, etc. Such measures have systematically and almost universally failed to ask several important questions: Did the use of resources for treatment make a difference? Has the client's condition changed since treatment? If so, how? These are but a few of the specific questions that Krapfl addresses in his chapter, "Accountability Through Cost-Benefit Analysis".

The second major problem in developing systems of accountability is that of the technology and operations necessary to make information systems valuable as tools for decision making. While computer hardware problems seem manageable, the problems of developing computer programs to both manage the large fund of information to be considered and perform such operations as cost and cost-benefit analyses, are indeed formidable. Not the least of these is the very fundamental problem of developing standard units of measure, so that meaningful comparisons can be made across treatment modalities. While these problems will not be solved quickly or immediately, a promising approach to their solution is discussed by Ciarlo and Horrigan in their chapter, "Outcome Measurement and System Modeling for Managerial Control and Accountability".

Finally, there are problems that must be addressed regarding the uses of evaluative research. Given the present state of the art, what can legitimately be said regarding the uses of evaluative research to create accountability? What considerations, reservations, and qualifications should surround attempts to use evaluative research findings to make decisions on how mental health or human service resources are to be allocated? In addition, what are the limits of the uses of data resulting from evaluative research? Finally, who *should use evaluative research to make program related decisions? These are some of the many concerns discussed by Levy in his chapter, "The Use of Evaluative Research in Creating Accountability".*

Jon E. Krapfl

Public agencies in the human services have only occasionally been held accountable for their performances. Rather, they too often have been held accountable for the maintenance of a treatment model and philosophy, with the assumption that if the model is translated into a program which is appropriately staffed, clients in the system will benefit. This has been particularly true in mental health, where the mental hospital has historically treated the construct of schizophrenia, rather than the immediate realities of a person's behaviorally defined problem.

Difficult questions are being asked of human service programs; by consumers, by legislators, and by some human service professionals. The questions are important in that they have to do with whether or not the programmatic use of public monies made any difference in the lives of people and communities affected by the human service programs. The questions are difficult because, in mental health, at least, arriving at definitions of outcome variables is conflicting, puzzling, and still in its infancy.

In the following paper, Krapfl discusses some of the problems in developing cost-benefit measures of human service programs. Focusing on different levels and types of cost analysis, Krapfl indicates how different types of cost analyses relate to the fundamental problem of creating a public reporting system that is likely to facilitate the development of a more competent system of human services.

While, at this stage of development, cost analysis techniques leave a lot to be desired in their thoroughness, perhaps, as Krapfl suggests, being thorough is not as important as being listened to.

CHAPTER 11

Accountability through Cost-Benefit Analysis

Historically, too little research in psychology has been concerned with the evaluation of effects of intervention strategies. Psychologists, primarily clinicians, who have conducted research into the impact of human service programs, such as those in the field of mental health, have usually been concerned either with some type of process data analysis, or with the empirical support for a theoretical position.

Two factors may account for the lack of evaluative research: Primarily, the research techniques in which clinicians have been trained, have been techniques for basic research rather than for program evaluation. Secondly, the problems dealt with in mental health have proven to be particularly intractable and difficult to define, thus the program evaluation research which has been done has frequently shown little, if any, effect for treatment methods. Therefore, there has been little reinforcement for further carrying out these forms of research.

At the present time we are moving from process to orientation in the evaluation of mental health programs. This increasing concern with the evaluation of the effects of our efforts, the specification of some ways in which this might be done, and the growth of methods leading to program accountability, will be dealt with in the present paper.

THE PRINCIPLE OF ACCOUNTABILITY

I should first like to define accountability and discuss the importance and implications of the concept for mental health programming. The term may not have originated with Leon Lessinger, a former Associate Commissioner of Education, but he is its most articulate proponent. In the Office of Education, Lessinger strongly pushed the notion of accountability in education and made strong moves to provide tools for achieving accountability, including systems approaches (Lessinger, 1970, 1971). Since Lessinger first articulated his position there have been strong moves in the direction of accountability in education and a great deal of writing has been done in this area. The journal, *Educational Technology*, devoted its entire January, 1971, issue to accountability in education. However, to date, it would seem fair to state that accountability is primarily a set of attitudes about the educational process, and accountability systems have not been implemented to any significant extent within the educational enterprise, though there are some signs of progress.

Lessinger, of course, was primarily interested in the adoption of accountability practices in the field of education. Nevertheless, the term has spilled over into other areas of human services. Lessinger (1971) provides something he calls a counter principle to the Peter Principle. The reference, of course, is to the book of the same name by Peter and Hull (1969). This counter principle is called the principle of public stewardship through accountability, and I believe that Lessinger's notion to combat the Peter Principle applies very well to mental health systems. The principle goes as follows: "Independent, continuous, and publicly reported outside review of promised results of a bureaucracy promotes competence and responsiveness in that bureaucracy." Many of the statements which Lessinger directs toward public education as explanation for a lack of accountability can also be directed at the treatment provided through various mental health systems. For example, mental health has a captive market with no competition. There are usually no, or limited, alternatives for treatment. Leadership is under the control of experts, presumably unchallengeable by the public, and there is no method of accounting for failure.

Accountability in mental health is going to mean that mental health professionals or specialists are going to be held accountable for the results of their therapeutic endeavors and intervention efforts; and presumably, are going to be rewarded, or not rewarded, depending on the outcomes which they produce.

Current and Predicted Practices in Mental Health Systems

The human service systems which appear least affected by notions of accountability are those human systems which are concerned with the implementation and delivery of services in the general area of mental health, mental retardation and disability. These systems have traditionally been held least accountable for the outcomes which they produce, probably because of public attitudes that little, if anything, of significance could be done for the retarded or the developmentally disabled or mentally ill person. The general public has been led to expect two things from mental health specialists. First, that they will keep problem people out of the public eye. Second, that when they do provide treatment, it will be very costly, and will consist of profound and incomprehensible activities. This presumed inability to make judgements on the activities of professional mental health specialists has led the general public to approve or disapprove funds spent for mental health for reasons devoid of real information about whether value for dollar was being received.

Even to this point in time there is little communication between mental health specialists or professionals on the one hand, and those individuals who are responsible for supporting the services of these professionals on the other, with the latter group being mainly composed of legislators, mental health boards, etc. Consequently, most budgets are incremental, (i.e., determined primarily by the amount spent in the past, and unrelated to effective output). Anyone who seriously entertains cost-benefit questions, or anyone who asks serious questions relative to outcome or objective achievement, is often declared to have an anti-mental-health attitude, if not given yet a stronger diagnostic label.

There are, however, a growing number of conflicts between professionals in the mental health retardation-developmental-disabilities areas and the public. And these conflicts are likely to continue and to become more pronounced for several reasons. First, those who are responsible for providing funds for mental health services, be they representatives of the federal government, legislators, or members of boards responsible for the operation of community mental health centers, are becoming increasingly sophisticated concerning the functioning of mental health professionals. These representatives of the public no longer stand in awe of someone who carries the title of psychiatrist or psychologist. Furthermore, they come from backgrounds in which, very often, the achievement of objectives is a familiar part of their working life.

As government support becomes increasingly more difficult to attain, and increasing amounts of local resources must be used to provide services, critical questions are going to be asked concerning the relative effects likely to be produced by the use of available resources. These changing situations all involve forms of accountability. Until now it has always been assumed that this kind of accountability was not possible in the human services. But the increasing sophistication of the general public, and their more intimate involvement in the affairs of human services specialists, the increasing limitations and restrictions on available resources coupled with increasing demands for them, and the general growth of government and public services and their associated increased costs, are going to continue to apply pressure for better value for each dollar spent.

However, specialists in the fields of mental retardation, mental health, developmental disabilities, and others, know little, often virtually nothing, of the language and processes of accountability. The programs in which these individuals are employed have, if anything, been anti-accountability and anti-evaluation, preferring instead a strong public relations approach. They have assumed that the general public was incapable of analyzing or evaluating their performance and have implied that they had no right to do it. These specialists have rarely taken seriously the notion of producing effective outcomes, as measured for example, in terms of people whom they were able to restore to effective functioning ability outside of mental hospitals. Rather, they have always been willing to engage in a process, usually theoretically based, which hopefully, and according to theory, should have a beneficial outcome. Consequently, we can see a fundamental and increasingly difficult problem in the interface of the general society and the professional mental health specialists.

The training of mental health specialists, by and large, leads them away from concepts of accountability. Loose terminology is rampant; belief in expertise is encouraged, and process, rather than outcome, is stressed. The entire superstructure of mental health is supported by assumptions of "good heart", and "humane treatment", but positive outcome is assumed by some, and not considered at all by others. There has been a general belief that, if we treat someone with behavior problems humanely he will improve, leading the values of the mental health professional to be concerned with human process rather than with humane outcome. Behaviorally, mental health specialists reinforce others, and are reinforced by others, for their activities rather than the outcomes produced by their efforts. Furthermore, they too often find anyone with a different set of values to be "inhumane", or say that they "don't understand".

Alexander and Messal (1972) have identified a number of differences between the mental health specialists and those who think in planning-programming-budgeting systems (PPBS) terms. [PPBS is one type of system used to build accountability into a system, and will be discussed later.] The first difference is that PPBS specialists assume that any program has a purpose which is commonly agreed upon. Those of us who are interested in mental health know that such an assumption is difficult to make. There are, for example, distinctions made between treatment-oriented versus prevention-oriented treatment. A second assumption made by PPBS specialists is that the outcome of mental health programs can be measured in some quantifiable or measurable terms. Again, specialists in the mental health area know that this is, in fact, rarely done, and many consider it impossible. The third factor, which Alexander and Messal identify, is that PPBS specialists assume strong centralized authority for programs. However, individuals who are not in agreement as to purpose or theory, and have few ways of measuring their effectiveness, are unlikely to be interested in creating a centralized authority since one man's guess about treatment is likely to be as good as that of another. Finally, the PPBS specialists assume that effectiveness will be evaluated according to productivity or outcome. Again, if one cannot identify objectives and, therefore, cannot measure them, he can hardly evaluate whether or not he has achieved them. Furthermore, we are often willing to hold the patient responsible for treatment failure, calling him "resistive", or "not ready for treatment".

AVAILABLE ANALYSES FOR DEVELOPING ACCOUNTABILITY PRACTICES IN MENTAL HEALTH

There are several approaches available for developing accountable systems in mental health. The two most important approaches are systems analysis and behavior analysis. The behavior analysis approach is most often focused on the performance of an individual within the system, on the specification and measurement of behavioral objectives for that individual, and on the design of intervention strategies and evaluation systems to determine whether or not the behavioral objectives are achieved. The behavior analysis approach has a strong scientific base, a rather large body of basic literature, and a well-developed methodology. When used to solve human problems, behavior analysis is most often identified as behavior modification, behavior technology, precision teaching, or applied behavior analysis. It is perhaps important to remember that this is the basic form of analysis for human behavior, which includes the performance of all of us, not just the patients and residents of institutions. Thus, we then employ behavior analysis to understand the behavior of the systems analyst, and, in doing so, may have a marked effect on what is called "Systems Analysis".

Systems analysis, the other major approach used to achieve accountability, differs from behavior analysis in its concern with the accountability of total organizations and systems. However, it often can and does employ behavior analysis to understand and control the behaviors of individuals within the system who are responsible for its functioning.

It would be appropriate at this point to provide a working definition of systems analysis or the systems approach, and I believe a most useful one is provided by Deterline (1971). He defines the systems approach as, basically, a methodology requiring exact specifications of what is to be accomplished by each component of a system, and the use of techniques to see that each specification is achieved. Meeting specifications is referred to as quality control, and seeing to it that specifications in quality control work, is accountability.

In systems analysis we are concerned with the specification and measurement of systems and organizational objectives and with the explication and evaluation of intervention strategies to determine their effectiveness in achieving system objectives. The terms most commonly used to describe the systems approach, or facets of the systems approach, are: system analysis, planning-programming-budgeting systems (PPBS), cost analysis, cost-effectiveness analysis, cost-benefit analysis, and cost-per-behavior-change analysis.

This paper is primarily concerned with the use of costing techniques such as cost analysis, cost-effectiveness analysis, and cost-benefit analysis in providing accountability in mental health. Though the distinctions between the various kinds of costing techniques are somewhat arbitrary, there are a number of different kinds of costing techniques and each of these techniques provides different information and, therefore, fulfills different functions.

Cost-benefit analysis is actually the most complex analysis of costs and has, as its basis, a number of different cost analysis techniques and other systems techniques which must be available before actual cost-benefit analysis can be carried out effectively. For purposes of discussion, I have classified these techniques into three levels of analysis which they provide. The categorizations are somewhat arbitrary and are not intended to appear mutually exclusive.

Descriptive Analysis

The first level of analysis can be called descriptive analysis. Descriptive analyses include such things as the ability to define successes in terms of desired behavioral or performance outcomes, and the inputs (costs) which are available for achieving these outcomes. Descriptive studies are post-dictive. They describe programs which are completed or are ongoing.

A major requirement for descriptive studies is the identification of explicit objectives which may or may not be achieved. This is the *sine qua non* for all forms of analysis since, without the articulation of objectives, all other forms of analysis are relating costs to an unknown. This simple notion of clearly articulating goals is a particularly difficult one in mental health. It is difficult to get common agreement among professionals concerning the form of these objectives. However, articulating objectives, and defending the objective which you have articulated, are two different events and should be dealt with separately. Anticipated criticism of selected objectives does not justify vague specification.

A somewhat more complicated level of descriptive analysis is known as a cost-per-behavior-change analysis, and includes not only the specification of behavioral objectives, but also a determination of costs required to achieve them. In this form of analysis one pinpoints as precisely as possible those behaviors which are required in order to state that an individual has achieved an objective. For example, in a mental hospital we may say that a patient must communicate with others on certain topics, must engage in a certain number of self-help skills, must maintain minimum standards of appearance, and so forth, prior to being dismissed from the hospital. The identification of each of these specific required behaviors, and the determination of the costs to generate those behaviors, is known as a cost-per-behavior-change analysis.

This level of analysis is the one most often neglected, even by experts, yet it is the most fundamental form of evaluative cost analysis, and can form the basis for the most accurate predictive and comparative studies. If we are interested in the effects of treatment programs, it is at this level that we should conduct important preliminary analyses, since people's behavior, broadly defined, is what mental health programming is about.

Another form of descriptive analysis, cost analysis, is a higher level of analysis, but one which still qualifies for a descriptive label in that it is a description of events which have taken place or which are currently taking place. In cost analysis we are concerned with the measurement of both qualitative and quantitative features of output from the mental health process, and the identification of costs for each output phase. For example, in cost analysis we are interested in identifying the number of institutionalized individuals who are returned to functioning in a work capacity outside the institution, or in the number of individuals who are able to return to community life, albeit without functioning in paying jobs. In cost analysis we may be interested in objectives which involve employing people in, or graduating people to, other forms of treatment or to other therapeutic environments where they can maintain already acquired behaviors or develop functional skills which could not be developed unless they were eligible for that environment. We may be interested in following individuals for long periods to determine the long-term effectiveness and costs of various intervention strategies.

It is important to recognize that cost analyses have objectives which are, in turn, composed of those specific behaviors (behavioral objectives) which were previously identified in the cost-per-behavior-change analysis. In cost analysis studies we should identify both the specific behaviors which must be changed and the resulting system objectives which will be attained by achieving all of the specific behavioral objectives. In addition, we should determine all of the costs associated with change in specific behaviors as well as costs of achieving system objectives which will lead to the patient's functioning adequately in a different (and improved) environment from the one in which he was treated. Thus, this type of analysis provides a measure of the cost of moving an individual from one stage of treatment to another or to the end of treatment. It should be mentioned that, for both of the descriptive categories already described, it is at least as difficult to define the behavioral objective in the instance of the behavior change analysis, and the system objective in the case of the cost analysis, as it is to describe the costs associated with the achievement of those objectives. However, attaining this information is extremely important, for without it predictive and comparative analyses are not possible.

Cost-Effectiveness Analysis

We move next to the predictive level of cost analysis in which we include those analyses which seek to establish costs for systems not yet in operation. At this level we might label the effort as a cost-effectiveness analysis. Cost-effectiveness analysis includes as inputs all of the costs identified in cost analysis as previously discussed in the descriptive studies. In cost-effectiveness analysis we attempt to *predict* the cost of achieving certain objectives and to define the strategies and resources required to achieve those objectives. In addition, we are required to convert all resources to dollar amounts so that it is possible to project budgets in order to achieve specific outcomes. It is possible, of course, to do a predictive cost-effectiveness analysis without prior input data (i.e., without prior cost analysis and cost-per-behavior-change analyses), but one must rely upon a logical analysis for determining predicted outcomes and predicted costs. However, it is precisely in this way that cost-effectiveness analyses have proven less than adequate. They probably should not be carried out until descriptive analyses are available and can provide some input data for predictive studies.

Cost-effectiveness analyses may also be used in a comparative sense in that, once one has identified the objectives to be achieved by a system it is possible to predict the costs of alternative implementation strategies, or to predict the relative output which will result from alternative intervention strategies. Thus, cost-effectiveness analyses may answer such questions as which intervention strategy is likely to produce the better outcome for the same funds, or which intervention strategy will meet minimum outcome objectives for the least amount of money.

A more sophisticated form of cost-effective analysis is cost-benefit analysis. The complexity of cost-benefit analysis, when contrasted with cost-effective analysis, is considerable. In cost-benefit analysis, we are interested in the comparison of the dollar value of the inputs required to achieve an objective with the dollar value of the output; that is, the *dollar benefit* which will result from the attainment of the objective (not just the cost of attaining the objective as in cost-effective analysis). It is a fundamental characteristic of cost-benefit analysis that every element with which it deals is converted to a common base, usually dollars. The difficulty involved in converting every input element and every output element into actual dollar value is the fundamental problem for the cost-benefit analyst. It is also the fundamental problem for the system purchaser, the individual or group for whom the cost-benefit analysis is being conducted.

Converting the achievement of the outcome objectives into dollar values is a serious problem in cost-benefit analysis. In mental hospitals, the dollar value of benefits gained is usually described with the assumption that the government is the systems purchaser. Thus benefits attained are usually expressed in the form of the amount of money the hospital saves through discharging a patient, and the resultant input of tax dollars (from his salary) into the system. These benefits gained are compared to costs (conceptualized as benefits lost), and the resulting ratio is called a cost-benefit ratio. If the cost-benefit ratio is greater than one, presumably the system purchaser should move ahead with the system since he cannot afford not to have it. If the cost-benefit ratio is less than one, then it costs money to have in operation the system being assessed.

It is extremely important to note that the cost input and estimates of benefit for all of the sophisticated costing techniques are taken from the basic cost data which results from descriptive cost analyses. It is the thesis of this paper that failure to explicitly identify the objectives which we are to achieve in mental health, and our consequent failure to identify the cost of achieving those objectives, means that more sophisticated analysis, such as cost-effectiveness or cost-benefit analysis, is rendered premature and inappropriate in that the latter are based on inaccurate and irrelevant input data.

COST-BENEFIT ANALYSIS IN HUMAN SERVICE SYSTEMS

In order to justify the assertion that higher level cost analyses are questionable, or even inappropriate prior to evaluative research, this section of the paper will be devoted to a brief history of cost-benefit analysis, and to the identification of problems associated with its use in human service systems.

Cost-benefit analysis, as a technique which provides decision makers with a method for systematic comparison of costs and benefits of alternative approaches in achieving outcomes, has been with us for some time — most frequently in the Department of Defense.

In the early and middle 60's there was a move to implement program planning and budgeting (PPBS) techniques, cost-effectiveness and cost-benefits techniques in the human services areas. In 1965 President Johnson issued a directive that extended the use of PPBS into most of the agencies in the executive branch of the government. While everyone could agree that effective systems analysis approaches would be quite valuable in the human services areas, many doubted that such an objective could be achieved. In general, the promise of these approaches was, and continues to be, that they would be able to specify clearly and precisely the objectives of the national programs which were being funded by the Congress, and that, of the various goals or objectives established, it would be possible to specify those which must be dealt with immediately as opposed to those which could be dealt with on a longer-term basis. In addition, the use of various methods of systems analysis were to provide a means of analysis by which alternative strategies for achieving outcomes could be compared with respect to their effectiveness, their costs, and their relative effectiveness for cost (cost-benefit).

For several reasons few individuals believed that the systems approach would be effective in areas other than those typified by the Department of Defense. First, the Department of Defense has traditionally had a great deal more power over its personnel than have most other agencies. There are clear and explicit chains of command and the commanders have authority over much of the life style of the individuals with whom they are concerned. Secondly, the systems approach, as employed in the Department of Defense, had been primarily involved in the analysis of hardware-oriented systems, rather than human service systems. Human service systems are more open ended and, therefore, more difficult to analyze. It has always been a good deal easier to work with machinery than to work with people.

Systems principles and cost effectiveness principles were employed in the evaluation of many of the poverty programs started during the Johnson administration. Cost-benefit analyses have been reported for several of these programs. For example, OEO conducted a cost-benefit analysis of the Job Corps which rather nicely illustrates some of the problems associated with the technique, especially in human service systems. As reported by Rossi (1972), three separate cost-benefit analyses were carried out on the same system. Analysts employed by Job Corps found a cost-benefit ratio of 4.0:1. An independent economist, evaluating the same system estimated the ratio at 1.5:1. The General Accounting Office generated a ratio of 0.3:1. Obviously when different analyses are carried out, and one of them finds a benefit ratio thirteen times as high as another, we need to raise serious questions about the utility of such analyses.

Cost-benefit analysts have a penchant for dealing with systems in all their complexities, probably because systems analysis and cost-benefit analysis were originally developed and used in the Department of Defense, where they dealt with complex issues on a very large scale. One rarely finds written evaluations of systems techniques or cost-benefit techniques applied to smaller units under carefully controlled conditions. And, because cost-benefit analysts and systems analysts want to be comprehensive, being concerned, for example, about the interface of a mental health system with other systems, most of the questions asked by the systems analyst or the cost-benefit analyst have little to do with the day-to-day behavior of the professional mental health specialists, except in indirect or summary ways.

When one realizes that the form of the analysis is highly theoretical and rests on a great many assumptions, a number of which may have no sound basis, one becomes aware that what emerges in many systems analyses or cost-benefit analyses of mental health, mental retardation, or developmental disabilities programming is likely to be highly inconclusive and perhaps of no practical assistance whatever. This occurs because what the systems analyst is doing is some very sophisticated juggling of some highly suspect and tenuous inputs. Thus, no matter how sophisticated the mathematical calculations or the logical analyses, the output measures can never be more reliable or more tenable than the input data upon which they are based.

In reviewing the problems of cost-benefit analysis as it might be applied to mental health programs, several features stand out. First, the studies tend to have been synthetic rather than analytic. And, we have an inappropriate data base for adopting models. As indicated earlier, planning and decision making on the basis of an analysis of what has transpired can only be as good as the input data itself, but we have lacked good input for a number of different reasons. For example, we have very poor design-effectiveness analysis; that is, we have not yet stated, in a functional way, what it is that we would like to accomplish through the use of cost-benefit analyses. We have not asked the question: "What consequence will result from carrying out this analysis?" Neither do we, as yet, have systems for effectively evaluating staff performance or assessing management effectiveness.

In a conference of the American Academy of Arts and Sciences Seminar on evaluation research (Rossi, 1972), a number of criticisms of OEO programs, which also seem to apply to the mental health movement, were identified. The seminar suggested that we have been confronted with vague goals, strong promises, and weak effects. To this we might add weak and insensitive measurement procedures.

It is impossible to conduct a cost-benefit analysis when goals cannot be stated in terms of desired outcomes. It is also very difficult to delimit the costs and benefits which should be considered in carrying out the cost-benefit analysis. It is this factor which appears to result in the marked variations in cost-benefit ratios which have been carried out in human service systems. The inclusion, or failure to include, certain costs and benefits lends a certain unrealistic quality to the ratio. For example, an individual discharged from a mental hospital and returned to the labor force pays a certain amount of taxes. Are these to be included under benefits gained? To what extent do resources provided for mental health cut into resources available for education? Should this be considered under "benefits lost" in the equation?

Additional difficulties are encountered when one becomes concerned about the costs of time — the discount rate which should be applied is very often unclear. However, this seems to be a rather trivial consideration since the margin of error produced by inappropriately identifying the discount rate appears to be relatively small.

Another difficulty encountered in cost-benefit analysis is the inability to assess the extent to which results are influenced by the experimental design which is used to conduct the study. Cost-benefit analyses are, by their nature, comparative analyses. The basic problem with comparative designs employing statistical analyses are the usual ones of: 1) equating an experimental group with a control group, 2) denying treatment to a portion of the population, 3) controlling extraneous factors which may account for change. While these questions are valid, and can be dealt with, doing so would require far more space than that allotted in this paper.

Furthermore, there are more telling arguments against traditional models of empirical experimental designs. The first is that any design which shows significant differences does not require statistical analysis. It will stand on its own. Demonstrating highly reliable trivial differences is a right reserved for basic science and has no place in a cost analysis. Secondly, statistical comparative designs are descriptive-evaluative in nature, and do not lead to control. In other words, they are only indicative of the present state of affairs, but provide no remedy. Designs which fulfill the latter requirement require continuous evaluation and will be discussed in the next section of this paper.

The use of apriori analyses (such as cost-benefit) before ex-post analysis (descriptive cost analysis) has already been posed as a problem. However, as a final consideration, we should look at other possible forms of apriori analyses which might be substituted for cost-benefit analyses, such as process-oriented qualitative research, for example, or justifications by appeal to authority. The basic problems with these approaches is that we have tried them and they simply don't work. In short, it appears that cost-benefit analysis or cost-effectiveness analysis is the only potentially viable form of demonstrating cost accountability in spite of its considerable shortcomings.

WHAT SHOULD BE DONE?

In discussing what should be done to achieve accountability through cost-benefit analysis, I would like to look, first, at the behavior of the cost-benefit analyst rather than cost-benefit techniques themselves. Cost-benefit analysts working in mental health, have been largely concerned with accountability practices within a mental health system. While no one would disagree with their objective of generating a more effective accountable system, it may be that an immediate focus on that problem is not the best way to solve it. In its place, I would suggest that the cost-benefit analyst should spend some time articulating the manner in which he is to be held accountable in the system.

Traditionally, cost-benefit analysts have seen their role as one of providing decision-makers with information which make better decisions possible. I would argue that such a role definition removes the cost-benefit analyst from an accountable role for himself. Thus, the Job Corps analysts could say that they generate cost-benefit ratios for the system, and the fact that they were not used to modify the system was the fault of the decision-makers.

A more accountable role for the cost-benefit analyst would be to set as an objective, changing the decision-making behavior of the decision-maker so that he uses systems data in making decisions. This seemingly slight difference between providing information to the decision-maker, and changing the decision-maker's behavior has profound implications for the cost-benefit analyst. It means that he must not only provide evidence for evaluating and improving a system, but must also see to it that the system is changed. The cost-benefit analyst thus becomes responsible for changing his own practices in order to achieve this objective. The output of the cost-benefit analyst becomes the input for the decision-maker. If that input does not affect the decision of the decision-maker it is of little or no value.

For the remainder of this paper, then, I would like to consider what the cost-benefit analyst must do when his objective is to generate analyses which have an impact on the system being evaluated. The discussion will be concerned with the following points which are suggested as relevant, but not necessarily exhaustive:

1. Identification of the system's purchaser.
2. Improving the validity and believability (read impact) of his output.
3. Design a program for evaluating his own effectiveness.

Identification of the System's Purchaser

The system purchaser is that agent or agency for whom a cost-benefit study is being conducted. In most instances he is the same person as the decision-maker, but this is not always the case. In fact, we are here introduced to the first dilemma in planning a cost analysis program for mental health.

That is, is the agent (system purchaser) a representative of society or is he an individual member of society? Let me make the distinction more clear. We would agree in general that public human services are responsible to society for their operation and effectiveness; it is ultimately society to whom they must be accountable. However, in no instance is any element of society directly responsible to the entirety of society itself. Rather, there is a responsibility to individual members of society or to its representatives. Since budgetary decisions in mental health systems are, by and large, made at the state level, we can assume in this instance that the system purchaser is state government, in the form of a governor and his executive office, the legislature, or more specifically, a committee or sub-committee which is responsible for the development and funding of mental health programs within a state.

However, we may also be responsible to individual members of society in the instance of mental health programming. This might be the patient in the hospital, or the family immediately responsible for the individual functioning within a hospital setting. Both of these groups — the representatives of society responsible for budgeting, and the members of society who are served by the system — might be conceptualized as the system purchaser. Thus, it is important to recognize that, as separate system purchasers, they may be interested in solving different problems or answering different questions. For example, the legislature may ask such questions as: "Given our desire to achieve a specific outcome, what will be the cost of that outcome?"

These are indeed very different questions to answer. However, certain facts are available which will help make these decisions regarding answers: First, it is unlikely that there will be any marked increase in funds available for mental health and mental retardation programming in the next decade. Second, "the attitude of the federal government and state government towards the provision of human services to the disabled and debilitated appears to be constrained at best." Given that it is the representatives of the people and the managers in their employ who will be making program decisions in the foreseeable future, it would seem that the cost study information should be directed to these persons. But it seems imperative, that when conducting costing studies, we keep in mind the interests of that smaller segment of society whom our services directly affect.

Improving the Validity of the Output

It is in this area that cost-benefit analysis has been most seriously deficient. A primary problem seems to be the level at which we conduct our original analysis. We provide cost-benefit studies of overall systems such as state mental health systems or studies on the national impact of community mental health centers. These studies are so broad in scope, require so many assumptions to be made, and have such unclear goals, that the impact on the system purchaser is usually minimal because of the variability of the data (cost-benefit ratio). This variability in data appears to be due to two factors: decisions as to what is to be included in the calculations of benefits, and the manner of determining cost (benefit lost).

With regard to benefits gained, the approach has typically been to include all potential benefits. For example, if it costs $10.00 per day to keep a patient in a hospital, and if we assume that we are dealing with a patient who is likely to be hospitalized for life, and if the patient is 51 years of age and her life expectancy is 61 years, then we can calculate part of the benefit of her restoration to society by multiplying $10/day by 365 days by 10 years, or $36,500. (Other factors, such as the discount rate go into the calculation, but are not germane to this point.) In addition to this we could include, as benefit, the taxes she will pay if restored to the work force for ten years, and also the sales tax on items purchased as a consumer, and so forth. A rationale for the inclusion of all these figures can be derived, and the intent of most systems analysis has been to be exhaustive in this respect.

However, I would suggest that being thorough is not as important as being listened to, and that the extent to which one includes such long-term potential benefits should be governed by the reaction of the decision-maker. I think that viable cost-benefit analyses can be carried out using only expenditures and potential expenditures within the mental health system itself. The argument here is that we evaluate programs in terms of their direct impact on the system under study and not in terms of the overall impact on the society. This narrowing of focus may be one of the ways to increase the credibility in the cost-benefit ratios generated, and consequently increase the likelihood that the decision-maker would use that information in making decisions. Whether or not more narrowly focused studies would have more impact on the decision-maker's behavior is, of course, an empirical question and should be dealt with as such.

If the objective is to increase the impact of the cost analyst's output on the decision-maker's behavior we should, perhaps, ask a more general question than how to improve cost-benefit analyses as such. The question should really be whether cost-benefit analysis or some other costing technique should be used to have impact on the decision-maker. Again, it is an empirical question, but I believe that a descriptive cost analysis of existing programs would be more useful at this point for two reasons. First, a decision-maker is more likely to be influenced by a careful analysis of something which is already taking place than by a hypothetical analysis of some potential event which will take place in the future. Secondly, good cost data on existing programs can provide a more ideal basis for making cost estimates of future programs.

Design a Program for Evaluating His Own Effectiveness

This third suggestion for what should be done now is the most important. To provide a detailed statement of the form of such a self-evaluation would be a paper in itself. It would include activities such as identifying criteria against which the value of a cost analysis could be assessed, and determining whether or not those criteria were met.

The form of the analysis is not as critical as is the adoption of a model which provides the cost analyst with a way of systematically monitoring and modifying his own activities as a function of their impact on a decision-maker.

In summary, cost analysis studies would not only provide an evaluation of current practices and projected benefit from future programming, but should also be designed and evaluated with regard to the extent to which they have impact on the system and are successful in getting the system to modify its practices. In order to accomplish this goal, the cost analyst will be required to scrutinize and modify his own practices until he attains the desired effect.

REFERENCES

Alexander, J. B., & Messal, J. L. *Hospital and Community Psychiatry*, 1972, Vol. 33, No. 12.

Deterline, W. A. "Applied accountability." *Educational Technology*, 1971, Vol. 11, No. 1, pp. 15-20.

Lessinger, L. M. "Robbing Dr. Peter to pay Paul: Accounting for our stewardship of public education." *Educational Technology*, 1971, Vol. 11, No. 1, pp. 11-14.

Lessinger, L. M. "The powerful notion of accountability in education." Paper presented at the Academy of Educational Engineering, Oregon State Department of Public Instruction, Wemme, Oregon, August, 1970.

Peter, L. J., & Hull, R. *The Peter Principle.* New York: William Morrow, 1969.

Rossi, P. H. "Testing for success and failure in social action." In Rossi, P. H., & Williams, W. (Eds.), *Evaluating Social Programs.* New York: Seminar Press, 1972, pp. 11-49.

Although the technology of evaluative research is still in its infancy, social and political pressures for valid and reliable measures of treatment-program effectiveness continue to mount. While, as the authors of the following paper note, it might be rationally argued that the emphasis on program outcome measurement and assessment should be no greater than existing and available technology, the realities of funding patterns are forcing, however prematurely, a critical examination of mental health treatment programs.

In the paper which follows, Ciarlo and Horrigan describe an experimental and highly innovative approach to treatment-program outcome measurement. Their developing system, a quantitative method for assessing program quality, holds considerable potential for the emergence of effective program management.

The past is replete with program development and termination because of the program's fit, or misfit, with particular theoretical models of treatment; in short, management decisions based on the ideological bent of the manager. If the future is to differ from the past, new technologies must be developed. Ciarlo and Horrigan provide a glimpse of one such method.

James Ciarlo

Jack A. Horrigan

CHAPTER 12

Outcome Measurement and System Modeling for Managerial Control and Accountability

OUTCOME MEASUREMENT, EVALUATION, AND ACCOUNTABILITY

Perhaps no single concern of mental health program administrators is currently more problematic than that of *accountability*. To many of these administrators, accountability in its strictest sense means *responsibility* — the responsibility of the program managers and staff to obtain positive end-products, results, outcomes, or output. Moreover, these results or outcomes must be assessable in clear, unambiguous, and, preferably, quantified terms. We share this definition of accountability primarily as the responsibility for results or outcomes; and we will try to show here how our Mental Health Systems Evaluation Project is relevant to this form of accountability.

Public programs are increasingly subject to scrutiny by governmental regulatory agencies, budget officials, legislators and their committees, and (as the "heat" comes on) the next higher echelon in the agency bureaucracy. Not only must waste, "fat", and inefficiency be reduced; the program must also "specify its objectives, and state how well the program is meeting them." How did we ever arrive at this uncomfortable state of affairs? Why is it that qualified mental health professionals can no longer be funded simply to provide the best services of which they are capable? Why must mental health professionals now become program analysts — defining goals, objectives, quantitative indicators of success, benefits in relation to costs, and the like?

We wish we could say that the new accountability demands grow basically from the fact of the availability of suitable evaluation technology, and that we are now fully capable of just this type of program analysis and program budgeting. But this is far from the case; there is currently no *outcome* evaluation or accountability system known to us which is adequate for what is being demanded. We have not yet systematized and measured mental health services outcomes reliably and validly. Nor have we been able to demonstrate (with few exceptions) the relationship between services and positive outcomes. Clearly the success of mental health research, evaluation, data-systems development, cost-accounting, and operations research have *not* been the impetus which has convinced our public leaders that the time is ripe for placing all our programs on an objective and factual footing.

If the above is correct, then the growing demand for placing the burden of program justification upon the ability to demonstrate positive results must stem from other forces outside the mental health system. Consider, for example, the ruckus over the question of the adequacy of our educational system during the early Sputnik era — a reaction better attributed to a perceived national threat than to evaluation of educational system effectiveness.

More recently, other social institutions have come under attack and have been sharply questioned as to motivation, adequacy, and capability — the medical establishment, the universities and their research institutes, our police and correctional system, and the social welfare and unemployment compensation system.

The roots of these manifestations of public disenchantment with long-standing social institutions may be difficult to specify. It seems quite possible, however, that they include *anti-intellectualism*, as opposed to a high valuation of theory and methodology in social programming; *anti-liberal ideology*, rather than true concern over the people's welfare; and insecurity about national values and goals, as contrasted to a solid consensus about our social values and political ideology as reflected in public programs. If this is, in any sense, true, then the demand for proof of positive outcomes in program justification, and the implication of swift funding cutbacks resulting from a lack of data illustrating positive program outcomes, may be much less rational than it seems.

How then should we view the present push toward accountability? Is it truly a manifestation of a demand for "social responsibility"; or is it primarily a facade which conceals basically irrational social and political forces of the times? It may not be possible to answer this definitively. However, we can assume a stance concerning accountability which is sensitive to both possibilities. We should resist the irrational components, yet encourage both the notions of public concern for people's welfare, and the allocation of resources into services which promise the best outcomes and greatest returns.

Most simply put, we can take the position that *there should be no greater emphasis upon outcome accountability than is consistent with the availability of adequate technology to determine true outcome.* Demanding more than this is akin to expecting to somehow get something out of nothing. While our position implies that true outcome-oriented program accountability must be postponed for a while yet, we must get on with the task of developing tools and methods of making such accountability possible. Moreover, control and funding agencies should insist that such efforts be initiated within the currently-funded programs and back up this insistence with adequate resources for the task.

We might go one step further and argue *against* the imposition of outcome evaluation upon programs by outside authorities before internal evaluation has begun. We would strongly prefer an accountability-oriented *internal guidance system* for program evaluation, planning, and decision-making as opposed to an externally-imposed set of accountability criteria and procedures.

Such internal systems might look very different at two different locations and might serve to raise the question of systems compatibility in comparing the two different programs. However, the gains of internal accountability might outweigh what we would lose in strict inter-agency comparability.

Perhaps the internal system would induce clinicians to provide more thoughtful services, and innovate more in-service techniques, rather than to resentfully wonder whether their efforts were going to show up as worthless. Perhaps the internal system would also be perceived as a valuable management tool, useful in helping make decisions about how and where to allocate the resources available to best alleviate the human problems that flow into a program. Furthermore, the internal accountability system could be seen as experimental itself, as much in need of continuing development and improvement as the program it is supposed to be monitoring and evaluating.

In our view, this type of internal accountability system is currently our only feasible option for fostering true "social responsibility" in public mental health programs. Such a system is less prone to misuse by over-zealous budget-cutters, and therefore less likely to support a rationale for dismantling possibly effective services merely on the basis of finance. It would probably evoke more responsible consideration of the consequences of fund cutbacks for the community being served. In addition, such an evaluation system could be exposed to public view, so that funding agencies could note whether program management was making the best use of a system that could be justified on the basis of its technical adequacy.

In sum, an effective internal system might function more as an instigator of progressive change than as an oppressive tool, and thus foster more true social responsibility than could be engendered via the threats associated with externally-directed accountability.

OUTCOME MEASUREMENT AND TREATMENT EFFECTIVENESS

At the Mental Health Systems Evaluation Project (MHSEP), we are attempting to construct and implement such an internal guidance system for a large urban community mental health center (CMHC). This Center annually sees about 12,000 clients, primarily at fifteen decentralized serving elements located within the catchment area. It provides a very wide range of classifiable treatments for these clients who exhibit problems of every conceivable type.

Initially, we group all of our clients into about ten major problem categories from which we can draw samples for follow-up in order to determine treatment outcome, and further analyze treatment effects. Our evaluation system is still in the developmental stage, but we are very close to a fully operational system for determining program outcomes and for instigating productive alterations in programs.

The basic component of the system involves a means of quantifying the actual results or outcomes for different types of clients seen. This is done through direct, face-to-face questioning and assessment of clients who have had any contact at all with the program under study. Our outcome measure is multi-demensional, partly because we set out to assess several important dimensions of client outcome, and partly because the questionnaire's item intercorrelations and scale-score analyses indicated that we were indeed measuring a number of orthogonal dimensions of client functioning (Ciarlo, Lin, Bigelow and Biggerstaff, 1972).

We usually refer to these dimensions by their negative or "pathological" poles: Psychological Distress, Interpersonal Isolation, Non-Productivity, Substance Abuse, Dependence Upon Public Support Systems, and

Trouble With the Law. We also assess Client Satisfaction with Center programs (including a client's own estimate of alleviation of his problems). Some of our outcome scales have excellent psychometric properties; others are marginal and are being improved. All have earned substantial acceptance by clinicians.

Briefly, we are attempting to do the following

1. determine the state of clients' functioning at some particular time following their entry into our service system;
2. relate the differing treatment careers of these clients to their subsequent functioning levels, in order to determine if any particular career or treatment technique is related to a more favorable client end-state; and
3. investigate what possible alterations in these treatment patterns within a given budget might improve the overall condition of the clients at follow-up, and/or what might be done to achieve the same average client outcome level for a greater number of clients.

A brief example of what we are doing is given below, in terms of a recent analysis of the Center's large alcoholism program. Table I shows preliminary scores on three of our outcome dimensions for three different samples of people: 1) a randomly selected sample of the Denver population in 1972, representing primarily residents of the Center's catchment area; 2) a sample of consecutive client admissions to the Center's alcoholism program in 1972; and 3) a sample of follow-up clients ten to twelve weeks after intake in early 1972. Higher scores indicate poorer condition or greater dissatisfaction with services.

The scores show that client condition at admission is much worse than the average reported condition in the community, and that at follow-up the clients' condition has improved, but is still substantially worse than the community average.

To check for the possible effects of time alone in accounting for lower follow-up scores, we compared the scores of persons receiving only a single evaluation (plus whatever "treatment" occurred at that same time) with those remaining in the program for at least one additional contact. Without control for possible client differences and motivations, the data suggest that treatment may not account for much of the difference between admission scores and follow-up scores for Psychological Distress and Substance Abuse. Thus, perhaps time alone could account for the observed differences, and we cannot rely heavily upon them to justify our program. Client Dissatisfaction, however, seems substantially and significantly lower in the group staying longer in treatment.

TABLE I

Mean Functioning Scores for Center Alcohol-Problem Clients and a Representative Community Resident Group

Dimension	Community Group (N = 89)	Admission Group (N = 29)	Follow-up Group (N = 156)	Test and Significance Level
Psychological Distress	4.07 (SD=2.02)	10.59 (SD=9.10)	7.83 (SD=6.59)	ANOVA, $p < 0.01$
Substance Abuse (Alcohol)	1.43 (SD=1.61)	6.55 (SD=4.34)	4.96 (SD=4.23)	ANOVA, $p < 0.01$

Dimension	Follow-up Group: Evaluation Only (N = 19)	Follow-up Group: Evaluation Plus Further Treatment (N = 131)	Test and Significance Level
Psychological Distress	9.32 (SD=6.23)	7.54 (SD=6.46)	t-test, N.S.
Substance Abuse (Alcohol)	4.26 (SD=4.02)	5.08 (SD=4.26)	t-test, N.S.
Client (Dis-) Satisfaction	9.59 (SD=4.65)	5.70 (SD=4.02	t-test, $p < 0.01$

Next, we tried to relate four different aspects of treatment to the outcome dimensions. To reduce the problem of non-random assignment to any of our treatments under analysis, we attempted to remove, via partial correlation techniques, whatever effects on follow-up scores were correlated with four types of client differences at intake — age, income level, ethnic group, and degree of living impairment (as rated by intake clinicians). The remaining partial correlations of the treatment variables with the follow-up scores are shown below in Table II.

In general, the correlations were not substantial; it does not appear that these follow-up scores are dependent to a great degree upon any of these four aspects of treatment career. Only two correlations were above .20 (note *), and we would need to find them again in our next follow-up sample of alcoholics before we accept them as reliable findings.

TABLE II

Partial Correlations of Treatment Variables with Outcome Dimension Scores for an Alcohol-Problem Follow-up Group (N = 150) After Influence of Certain Client Characteristics has been Removed

Treatment Variable	Psychological Distress	Substance Abuse	Client (Dis-) Satisfaction
Intensity of Treatment (Group output = 1, Individual Output = 2, Day Care/input = 3)	.11	.11	–.08
Length of Treatment (Scored 1 to 4 depending upon number of days in input/Day Care & number of Output Contacts)	–.06	.12	–.23*
Type of Serving Clinician (Primarily Paraprofessional = 1, Mixed = 2, Primarily Professional = 3)	–.09	–.02	.13
Use of Adjunctive Medication (Tranquilizers) (No = 0, Yes = 1)	.15	.32*	–.13

Some clinicians who saw these results expressed concern that the two largest correlations resulted from our inclusion of severe alcoholics, treated by the Detoxification Ward, in the total group of alcoholics studied. We therefore repeated the analysis, this time of only those alcoholics treated on an outpatient basis by the decentralized neighborhood teams. The results of this analysis are shown below in Table III. The two largest correlations reappear, and they are even larger than before. It is also worth noting that length of treatment now appears to be consistently related (although very weakly) to all three outcome measures.

These results suggest the possibility that: 1) by cutting down the amount of medication given to alcoholics, the program might be more successful, and 2) it might improve by trying to keep alcoholics in treatment longer. These changes, if instituted by management, and followed by subsequently better overall results, would tend to confirm these hypotheses which we are inviting program managers to consider. Is this non-experimental type of data strong enough to induce program

TABLE III

Partial Correlations of Treatment Variables with Outcome Dimension Scores for an Outpatient Alcohol-Problem Follow-up Group (N - 45) After Influence of Client Characteristics has been Removed

Treatment Variable	Psychological Distress	Substance Abuse	Client (Dis-) Satisfaction
Intensity of Treatment (Group Output = 1, Individual Output = 2, Day Care/input = 3)	.13	–.11	.08
Length of Treatment (Scored 1 to 4 depending upon number of days in input/Day Care & number of Output Contacts)	–.30	–.14	–.30
Type of Serving Clinician (Primarily Paraprofessional = 1, Mixed = 2, Primarily Professional = 3)	.12	.10	.22
Use of Adjunctive Medication (Tranquilizers) (No = 0, Yes = 1)	.20	.39	–.12

alteration or innovation? We are hopeful that it is, and we are studying managers' reactions to such feedback to find out.

An additional inducement to program alteration would be the presentation of hypothetical program results which predicted substantial outcome improvement for any given program change. The generation of such predictions is the primary function of the operations research phase of our project.

Operations Research

Operations research is involved in the MHSEP on two levels. At one level operations research ties the whole CMHC together, as a managed system, with a single model or series of models. At the same time operations research models parts of the CMHC of specific interest.

As described previously, our basic evaluation effort involves making observations regarding demographic, clinical, treatment, and outcome variables operating in a large urban community mental health center. Currently there are about 150,000 client contact records representing one-and-a-half years work by the CMHC staff. Nearly 1000 clients have been followed up and have outcome scores. Our basic evaluation effort has now settled into a rather steady operational status.

Our operations research unit is currently determining the costs of selected treatment careers for all clients served in the Center. Building upon the basic evaluation effort, the operations research team is also attempting to develop conceptual models and verify these for certain Center programs at particular points in time. While there are several types of models which can be constructed, we are beginning with linear models.

Looking for Functions

In order to do modeling at the CMHC, we must develop functions of at least two types: 1) outcome functions, and 2) operating cost functions. The raw outcome data are client "scores" to be related to client characteristics, treatments, and clinician/serving element data. The outcome functions may include consideration of the 25 or so different operating units in terms of unique and common operating characteristics as well as differences in serving element locations and client populations.

Raw cost data are gross estimates of cost as a function of service type performed. While it would be desirable to have costs broken out into functions of services performed by clinician types for specific client problems, we do not yet have such refined cost estimates. As we gather more data in each of these functional areas we do expect to be able to focus on the finer details.

Some correlation of outcome scores and clinician/client/service data has been undertaken (as reported earlier in this paper), but much remains to be done. When we know how outcome is affected by the various input and processing variables, we can then do more meaningful modeling.

We are faced with a number of unanswered alternatives: It may be the case that outcome is not the function of the known input variables. If this proves true, we might need to change our methods of measuring and recording data in order to capture important variables. Possibly, the Center might need to change client processing, or perhaps we may find that processing of any type is ineffectual. In the last instance, that would preclude modeling of any dimension.

However, over a span of many years, workers have observed that clients do ultimately get "better" given time and a selection of treatment regimens. Our observations to date strongly support this conclusion, and accordingly we are looking for the functions that tie outcome to effort expended.

As we discover definite functional relations we will be able to identify the processing functions which prove to be the most effective. It is our intent to recommend increased (or decreased) emphasis in service programs on the more (or less) effective processing functions. The apparent negative effect of chemotherapy on alcoholics (noted above) is an example of just such a process which we may recommend to be curtailed.

At the present time our main tools in the search for functions are partial correlation and multiple regression. Our data base includes a wide set of client and clinical characteristics such as age, sex, income, location of residence, race, presenting problem, diagnoses, and severity of impairment.

Also included are records of processing information, such as number of days of hospitalization or day care, number and length of individual and group contacts, serving clinicians, and medication. The follow-up outcome measure provides the dependent variables to which the preceding data is related.

Preliminary work clearly indicates that no single input variable will account for complete change in outcome that occurs. Of course, this is not an unexpected result. It does, however, mean that we are going to have to select and verify our modeling functions with care.

Linear Models

Our current operating assumption is that there are linear functions that express outcome as a function of the processing and input variables, and that costs may be expressed as linear functions of processing variables. It is quite possible that some variables may not be linear. If we identify any clear non-linearities, we will apply techniques suitable to handling them. Linear functions do guarantee rapid convergence to the solution of problems which have solutions.

There is a considerable amount of literature on linear programming models and many readily available computer programs for solving such problems. We have run some preliminary models with hypothetical data, as shown in the example on the next page.

Let us presume that we have 300 clients who have gone through five types of process modalities:

Modality 1 (M1) equivalent to 3 days hospitalization
Modality 2 (M2) equivalent to 5 days in day care
Modality 3 (M3) equivalent to 5 individual contacts
Modality 4 (M4) equivalent to 6 group sessions
Modality 5 (M5) equivalent to evaluation only

The costs for each might be:

Cost M1 = $400
Cost M2 = $250
Cost M3 = $100
Cost M4 = $ 60
Cost M5 = $ 20

The outcome scores (now scored in the *positive* direction) could be:

Outcome M1 = 27
Outcome M2 = 24
Outcome M3 = 22
Outcome M4 = 20
Outcome M5 = 15

If we had the total of 300 clients distributed:

Clients in M1 = 10
Clients in M2 = 20
Clients in M3 = 100
Clients in M4 = 100
Clients in M5 = 70

we would have a total cost of $26,500 and a total of 6000 in outcome scores. The question is, can we do better?

We can write the above in a shorthand notation:

$$\begin{array}{rrrrr} 400, & 250, & 100, & 60, & 20 \\ 1, & 1, & 1, & 1, & 1 \\ 27, & 24, & 22, & 20, & 15 \\ \multicolumn{5}{l}{\leq 26{,}500 \geq 300} \end{array}$$

The first line is the cost per client in each modality, the sum of these costs to total less than or equal to the $26,500 of the last line. The second line indicates that the sum of the number of clients is to be equal to or greater than 300, the second number in the last line. The third line is the outcome value for each modality.

We then call a program SIMPLEX, available on a local computer timeshare system, and submit this as a problem. SIMPLEX tells us to put

1325 clients into modality 5 (evaluation only), giving a total outcome score of 19875. While not exactly what we expected at first, this is not an unreasonable conclusion. If evaluation is that effective, we should do a lot of it.

However, real outcome for evaluation only might not compare as favorably with outcome for other modalities as is implied above. Therefore, we tried a new set of outcome scores, with a lower relative score for evaluation only:

$$\begin{array}{rrrrr} 400, & 250, & 100, & 60, & 20 \\ 1, & 1, & 1, & 1, & 1 \\ 30, & 25, & 20, & 11, & 4 \\ \multicolumn{5}{l}{\leq 26{,}500 \geq 300} \end{array}$$

Our cost for the original distribution of 300 clients would still be $26,500, but we now have a total outcome score of 4180. We call SIMPLEX again and it tells us to place 256 clients in modality 3 (individual contact) and 44 clients in modality 5, thereby giving us an outcome score of 5300. In this case, we maximize outcome using *two* modalities.

Let us readjust hypothetical outcome scores again, so that the outcomes are proportioned to costs:

$$\begin{array}{rrrrr} 400, & 250, & 100, & 60, & 20 \\ 1, & 1, & 1, & 1, & 1 \\ 80, & 50, & 20, & 12, & 4 \\ \multicolumn{5}{l}{\leq 26{,}500 \geq 300} \end{array}$$

The total cost for the original distribution of clients is still $26,500, but the total outcome score is now 5270. We call SIMPLEX and get back a set of solutions:

modalities	1 & 5	2 & 5	3 & 5	4	5
number	54 & 246	89 & 211	256 & 44	442	1325
total score	~5300	~5300	~5300	~5300	~5300

Now, any one of the five "solutions" is equivalent to the other four since each has about the same total score.

What do these hypothetical variations in the parameters suggest? Since the individual outcome scores in the first problem are more likely ones, at least according to our preliminary findings, we should recommend that management carefully reconsider alternatives to evaluation only.

Theoretically, the cost of evaluation could be raised, or the "value" lowered, but this is not reasonable. Shorter hospitalization and more outpatient work would reduce the cost of the usual course of inpatient treatment and thus possibly increase the total outcome value. Therefore, we certainly should recommend that management *not* use hospitalization as an *alternative* to outpatient treatments, unless hospitalization is absolutely required.

Obviously, more effective treatment modalities would also improve outcomes; therefore, development of new, more effective treatments should have a high priority within mental health systems. Shifting resources among clinician categories and changing emphasis on treatments could also possibly change outcome values favorably.

Some of these issues have been considered in our hypothetical analysis above. We do not yet have definitive data to provide precise input to analytic processing routines; however, the conclusions reached above would definitely suggest program alternatives for Center management *if* the input data were actual figures representative of our system operations today. When we do have reliable, verified information for modeling, we expect to provide center management with a set of conclusions *and* possible alternatives for action (should action be called for).

CONCLUSION AND FOLLOW-UP

The data we have presented is either preliminary or hypothetical. After we have generated actual figures, our next steps, with respect to the alcoholism program example discussed above, will be as follows:

1. Follow up another sample of alcoholics. Repeat the partial correlation analysis shown in Table II. If the results support the first set of findings, continue with the next steps below; otherwise, search for other outcome and cost functions.
2. Calculate the outcome scores expected from each client's length of treatment, using the regression equation. Calculate the costs associated with differing lengths of treatment. Using the linear programming model, determine what total outcome increment would be obtainable within current resource limits if treatment length were increased.
3. Determine the Substance Abuse outcome score expected for clients as a function of the use of medication. Calculate the outcome increment that might occur if medication were reduced to the minimum feasible level. Calculate the potential cost savings.

4. Calculate the potential cost savings if non-professional clinicians were used to the maximum in outpatient contacts, within the limits of lawful dispensing of medication and prudent personnel management policies.
5. Calculate the potential cost savings if hospitalization were reduced to the minimum deemed necessary by clinicians. Determine the trade-off potential of reduced hospitalization with increased length of outpatient treatment, by shifting resources from inpatient to outpatient care. Alternatively, determine the trade-off of reduced hospitalization with the treatment of a greater number of the community's alcoholics.
6. Feed the findings back to program managers and staff. Look for possible improvement in the next group of alcoholics who are followed up, and try to determine whether this improvement is a function of service changes. Monitor operating costs for significant changes

We should note again that none of our findings will be definitive, in the sense of carefully-controlled experimental research. It is our belief, however, that much can be accomplished in the usual clinical setting with the data-gathering, analysis, system modeling, and feedback activities we have described here. It is our hope that we will be able to provide the Center with a range of promising program alternatives, some of which may then be implemented to increase the overall effectiveness of its programs.

REFERENCE

Ciarlo, J. A., Lin, S. F., Bigelow, D., & Biggerstaff, M. "A multi-dimensional outcome measure for evaluating community mental health programs." Paper presented at the American Psychological Association Convention, Honolulu, Hawaii, September, 1972.

Public policy, politics, and representative community participation are all intimately related to the problems of creating accountability in human service programs through the effective uses of evaluative research. If, as Levy notes in the following paper, research is to mean anything at all beyond being an interesting academic exercise, there must be clear commitment by policy making bodies to the use of evaluative research results in implementing public policy. The composition of the membership of these groups, and the definition and extent of their personal and organizational commitments to the uses of evaluative research, become significant political problems for the program developer, evaluative researcher, and, perhaps most of all, the consumer.

Professionals in mental health, historically, have been known for their commitments to theories, whether or not the theories seem to work when translated into operational mental health delivery systems. Too often the evaluation of the effectiveness of a program has been on the basis of criteria generated within the same theoretical model or delivery system. Only occasionally have criteria external to a treatment system been used to evaluate its effectiveness. Even more rare has been the use of evaluative research based on those criteria in establishing social policy. In the following article, Levy poses some considerations which will help accomplish these tasks. While it is true that without adequate methods and techniques for evaluation we have no evaluation at all, it is equally true that without ways of effectively using evaluative research for the public welfare, the research, no matter how well done, means nothing.

Leo Levy

CHAPTER 13

The Use of Evaluative Research in Creating Accountability

EVALUATION — SOME PITFALLS ON THE ROAD

In any consideration of the issue of evaluation of social programs, one should take a firm stand on the matter of clear, specifiable program objectives. This position should require that in order for a program to be implemented, a list of objectives must be put forward. Further, these objectives must be operationally defined in terms of outcome criteria which are, or will be, readily available at that point in time at which they are needed. Beyond this, as Rossi (1971) has argued, there must be commitment on the part of program administrators, funding agencies, and consumers alike to abide by the results of evaluation studies which indicate whether, and to what extent, the stated objectives are reached. Too often, negative findings are either ignored or subjected to extensive rationalization and, in this manner, programs continue long beyond the point at which they have evidenced their ineffectiveness.

One must understand that all programs put forward to ameliorate a social problem have face validity, which is to say that they look like they are likely to produce the desired results. If a program did not appear ostensibly plausible it would not generally be undertaken. Yet we have a multitude of examples of eminently plausible, well-conceived applications of generally-accepted social science theory to the solution of a practical problem, which when subjected to evaluation, prove ineffective. For example, in a well-known study by W. B. Miller (1962), a total community approach to the reduction of juvenile delinquency consisting of a complex program of activities with local citizen's groups, with professional agencies who deal with adolescents, with psychiatric casework for families with histories of repeated and long-term utilization of public welfare services, combined with intensive street-corner-gang work, produced negligible results. Miller concludes (1962), ". . . All major measures of violative behavior — disapproved actions, illegal actions, during-contact court appearances, before-during-after appearances and project control group appearances provide consistent support for a finding of 'negligible impact' . . ."

Another example of this sort is the recent, much discussed and debated, evaluation of the OEO project "Head Start". In brief, a well-conceived study by the Westinghouse Learning Corporation (Cicarelli, 1969), produced essentially negative findings with regard to any lasting impact of Head Start programs on disadvantaged children. The report provoked a storm of criticism, and illustrates very well some of the practical difficulties created by this type of research. These types of criticisms are cited and effectively answered by Evans (1969).

Not all researchers are convinced of the necessity of stating specifiable outcomes in evaluating social action programs. Weiss and Rein (1969) take this tack and raise the argument that in many cases what we hope to ultimately achieve is too complicated to be measured by simple, direct, operationally-definable criteria. I believe that arguments against outcome studies are suspect, and a hedge against clear analysis. When no outcome variables for a program are stated, this renders the program incapable of failure (or success, for that matter). Weiss and Rein's plea for ". . . a more qualitative process-oriented approach . . ." boils down to a request for more program description which, while necessary, is certainly insufficient, in itself, as an approach to program evaluation. I'm afraid that should we adopt their approach, we would end up like the practitioners of Freudian psychoanalysis, with a process which is minutely and lovingly described but still seeking (after almost a century) the answer to the question, "Does it work?" The question is rendered meaningless because we cannot state in specific terms what psychoanalysis is supposed to accomplish.

Raising the subject of psychoanalysis leads to a peripheral problem concerning the complexities and vagaries of the process of human motivation. Programs are brainchildren of people who love them and are loathe to see them criticized, let alone abandoned. Beyond this, statements of desired outcomes or objectives of a program may vary depending upon the perspective of the person who espouses them. Desired outcomes may be unstated because of their socially unacceptable nature. An administrator may have as an objective the maintenance of his and his co-workers' jobs, which might end if a program closed down. A community might like the idea of funds flowing into it regardless of the program to which the funds are attached.

The general gist of the above argument is that the problematic nature of evaluative research, as applied to complex programs of social action, stems not so much from inherent scientific problems, but rather from a number of other social and political human complications which make the pursuit of normal scientific procedure especially arduous in this area of investigation. This is true, however, in other areas of scientific work. Scientists get to love their theories and are loathe to give them up — a state of affairs which prompted the ascerbic observation by a colleague, "Old theories never die, they just rot and stink forever."

The parallel between administrators and their programs, and scientists and their theories is apt. Both must be on guard against emotional involvement with their subject. This *caveat* is formalized in science by the ritual use of the null hypothesis which, contrary to normal human inclination, states that the relationship found between two variables is expected to be due to chance and it is the scientist's job to disprove the null hypothesis. Although this makes for an awkward conceptualization, it serves as a constant reminder to the experimenter that he should expect to be wrong in his beliefs. One might even put the case more forcefully and say that the scientist should do everything in his power to prove himself wrong, and only in failing to do so, consider his hypothesis supported (i.e., the null hypothesis disproved). A goodly dose of skepticism with regard to our pet programs aimed at alleviating one or another social problem would be a useful, and (in view of the past performance of behavioral scientists in general, and mental health professionals in particular, in this area), a more plausible stance.

The commonly assumed defensive attitude toward evaluation represents in large part our emphasis on training in techniques and expertise in rendering services, rather than on the matter of establishing truth or discovering regularities in nature. Professionals would do better to see themselves as dedicated to the pursuit of objectives rather than the exercise of specific techniques which may after all prove to be invalid or after a time, superannuated. It goes without saying that we must also discover a better way of subsidizing programs than to make people's jobs contingent on the success of an untried and unproven program.

EVALUATION — THE EXPERT VS. THE LAYMAN

Some years ago, while testifying before a congressional committee investigating urban problems (Levy and Visotsky, 1967), I received a shock. Another psychologist addressing himself to the topic of community responsibility in urban planning suddenly confronted the committee with the following proposition: "How can community people plan when they don't know systems analysis?" Discussion of this point led, as I feared, to the conclusion that this accomplished person really believed what he said and had not, as I had hoped, merely been carried away by the ecstasy of the moment and thereby led into what was charitably, a *non-sequitur*, or viewed in a more sinister light, an attack on the vital heart of the democratic process (i.e., community participation in determining its own destiny).

I am continually impressed by the pervasiveness of the technocratic stance not only among professionals, who after all make their living as experts of one sort or another, but among politicians as well. My home for the past decade has been Chicago, whose mayor of many years, Richard Daley, has been a proponent of technocracy, *par excellence.* In common with many politicians, he views himself as the community's choice of leader and spokesman. This is not unreasonable, since he has been elected, and several times re-elected, always by a sizable margin. He then presumes to speak for the city of Chicago, all of his appointed commissioners, and through his mandate, the community. With this mental set, it is no doubt vexatious to be confronted, as he and his commissioners repeatedly are, by citizen's groups demanding a role in policy-making for their neighborhoods, and, at times, for the city as a whole.

I have served for most of my Chicago years as a member of a professional advisory committee on mental health to the Chicago Health Department. There also exists a parallel committee of community representatives (nonprofessionals) who serve in a similar advisory capacity. Both of these committees have been in continuous altercation with the health commissioner (since my arrival on the scene, a succession of three) in connection with having their advice accepted. One particular sentiment emanating from both groups, and met with responses ranging from coolness to table-thumping hostility, is that the Health Department give communities greater authority in running their own community mental health centers — of which 18 exist at present.

The answer is always the same. It begins with the fact that the mayor is the elected representative of all the people of Chicago (not, by implication, the representative of one or another special interest group of which the present committee is an example). The explanation continues to the effect that the Health Commissioner is the Mayor's appointed representative and, as such, bears the administrative responsibility for the health of the citizenry of Chicago. Now this is an awesome burden, requiring not only special professional expertise, but special information (concerning a broad variety of matters including high finance and politics) to which only the higher echelons of the city administration are privy. At the end of this disquisition (it is sometimes very long and passionate), if it has the desired effect, one feels that he has indeed acted in a presumptuous fashion and should be totally ashamed to have entertained thoughts of meddling in such a delicate and intricate matter as planning mental health policy.

The attitude of the Chicago city administration has consistently been that it is the state's responsibility to provide mental health services. Thus, the municipal share of the budget for the city mental health services has been kept at the lowest figure possible. It has been raised year by year since 1960 only in response to public pressure funneled principally through advisory bodies. The state mental health department has done its part by demanding that its heavy input of funds to the city mental health centers be matched by the city, *and the city's dollars spent first.*

The benevolent, paternal stance struck by the municipal health department advisory committees at budget time has been effective.

The argument that "ordinary people can't plan", was picked as an example because I, at least, view it as extreme and, in its own way, humorous. The argument was repeated above in terms of city politics where it was stated, perhaps for some, more plausibly. The fact is that it is a very common argument and represents, in the final analysis, a search for power and privilege by, and for, an elite group with special, sometimes mysterious, credentials. Communities, when organized to react at all, tend to read this argument correctly and react with hostility to social scientists, for example, and other outside elite groups such as university-based experts, who approach them with plans for the community or, as they might say, put "designs on it".

An alternative view of people and decision-making is furnished by our legal process in the concept of the jury system. In a trial, expert witnesses are used, but evaluation of evidence and arrival at a verdict based on this evaluation is a lay decision made by twelve ordinary people sitting in judgement of a peer accused of a crime. In this situation, highly complex decisions are reached by persons meeting certain minimal standards of human competence. Just how complex this task is, is evidenced by the frequent clash of expert opinion which somehow has to be sorted out by the jury. When this protean democratic process is subverted as in some civil commitment hearings, for example, violations of an individual's civil rights become inevitable. Imagine a legal expert stating before a congressional committee that the jury system should be abandoned because ordinary citizens are not trained in the law — or for that matter suggesting that Congress itself should not legislate because senators and representatives do not have diplomas in the diverse areas wherein they legislate!

One deficiency with the democratic process, in comparison to an authoritarian system, is its inefficiency in problem-solving. It is in recognition of this fact that democratic governments veer sharply towards authoritarianism during crises. A too-literal application of the principle of participatory democracy has rendered an organization like Students for a Democratic Society (SDS) largely ineffectual as a political force. By contrast, the city of Chicago is not infrequently cited as one of the more orderly and efficient large city governments in the United States.

Totalitarian systems are ugly from the point of view of their victims. But from the point of view of their beneficiaries, they have advantages of orderliness, efficiency and relative safety. The United States has consistently supported certain totalitarian regimes in foreign nations on these and similar grounds. Democratic systems are noisy. Without the absolute authority of a leader to boom out the command to cease debate, debate is likely to go on forever. The German poet Heinrich Heine is reputed to have remarked, "If the world comes to an end I shall go to Holland — there everything happens fifty years later." While the remark was no doubt intended as a negative reflection on the Dutch character, the Dutch don't appear to take it that way. The Dutch are a profoundly independent and individualistic people who quarrel and discuss at great length. This tends in part to slow down action — or as some would have it, "progress". Things take time in the Netherlands, in large part because it is a democracy where every voice is entitled to be heard and no individual is regarded as too small to be counted.

THE ISSUE OF ACCOUNTABILITY

The preceding discussion has some bearing on the issue of accountability, which generally asks the question: what person or institution is kept honest by what other persons or institutions? In the present context those accountable are mental health professionals, and the institutions and agencies they staff have to do with mental health programs. The services which they render are directed to behaviorally deviant and emotionally disturbed persons, including entire communities and certain high-risk populations residing in those communities. One might ask several kinds of questions about these services:

1. Do these services do harm?
2. Do these services accomplish what they purport to?
3. Are these services the least expensive way of achieving the desired effect?
4. Is the degree of benefit achieved through these services commensurate with the cost?

A variety of persons and institutions have a right to pose these questions and to expect forthright answers: certainly the consumer, whether conceptualized as a person with an emotional problem or as a community of persons to whom an institution is addressing its ministrations; the varying levels of government which feed funds to the institution; the professional mental health community at large, who have the moral obligation to police their own activities; all have a right, if not the obligation, to pose these questions.

Typically, questions of efficiency (costs and standards) and harmful practices (malpractice and unethical behavior) are policed intraprofessionally. Professionals are understandably less concerned with effectiveness, priorities, indirect social costs and cost-benefit ratios. These latter issues become the concerns of persons who are either recipients of service themselves, or those who represent persons or communities which are the recipients of service from the professional mental health agency. It is the consumer of service and the agency which provides funds who want to know if a service is any good or if the benefits are worth the cost, or if spending money in one way is better than spending it in another. Professionals tend to assume that their services are effective. To them, not only are their techniques valid, but they are also worth their cost.

Nor are social priorities the professional's strong suit. The professional tends to do his thing for those who want and can pay for it. In spite of a good deal of public posturing to the contrary, professional persons who function in the context of a marketing culture behave much like other purveyors of goods and services in a competitive economy.

From the professional's point of view, evaluation is most often seen as a matter of monitoring technique and setting standards. He may ask questions relating to number of psychiatric beds per 1000 population, or about staffing patterns of mental health centers, or whether generally accepted techniques of therapy are being practiced, or whether the techniques are being practiced with a high degree of technical skill. The professional will also concern himself with matters which may affect his public image; such as, whether it is proper behavior for a psychotherapist to have sexual relations with a client, or does a therapist charge outlandish fees or accept kickbacks.

Is a Program Harmful?

Individuals and communities have the right to expect that ministrations by professional mental health personnel will not injure them or aggravate their condition. They may also expect that they will not be manipulated for the gratification of the professional. They have a right to expect a certain level of concern and performance. Certain practices, if known, are easily monitored and judged by lay persons. Therapists who seduce their clients, and administrators who promise services which are never provided can be held accountable, provided the necessary information is available. The problem here is that some of what goes on in a mental health operation (principally, privileged communications between a client and a therapist) is genuinely inaccessable to lay people. However, all other information concerning finances, programs, professional salaries, objectives and aggregated statistics on patients should be available as a matter of principle, to the public. Otherwise private interactions of clients with their therapists should be available to designated professionals for a system of peer review. In this manner, (advice of a disinterested professional advisory group) harm is usually easily detectable by the community.

Do the Services Accomplish What They Purport to?

Here, obviously, the matter of defining suitable outcome variables is crucial. These indices must be objective and consensually agreed upon by the professional staff and the lay board. Clear expectations for new programs which include not only end objectives, but intermediate targets as well, should be formulated. A hypothetical model of expected change in the outcome indices should be projected. Finally commitment to act on the results of the outcome study should be secured from consumer, professional staff, and funding agency, alike.

Is This Program the Least Expensive Way of Accomplishing the Stated Purposes?

Having determined that a program is successful, we may then turn our energies to the task of making it maximally efficient. This generally revolves around the issue of testing the elements of a complex program to see to what extent each sub-program contributes to the effectiveness of the total program. For example, had Miller's (1962) program in combating juvenile delinquency worked, one would then still have the question of the contribution of each of the elements to the total result. That is, local citizen action programs, professional agency programs, family casework, and street-corner-gang intervention constituted four elements which, if taken together, produced the desired results, might conceivably produce similar results in combinations of one, two or three elements. Research designs to test the efficiency of a set of discreet program elements are not hard to conceive, but as a practical matter, they may be difficult to execute.

Is the Degree of Benefit Commensurate With the Cost?

This issue (cost-benefit analysis) emerges logically out of the successive considerations of effectiveness and efficiency. It is conceivable, however, that a program may be effective and at the same time the least expensive (known) way of dealing with a problem; yet, the cost may be considered too great. The question of priorities enters here, and in the end is always a matter of relatively subjective choice. This choice should be made by the community-consumer, through a lay board and based on the following considerations:

1. If the choice is left to the professionals, it is reasonable to expect that the choice among programs will be made in line with the interests and competencies of the professional staff, regardless of the needs of the community-consumer;
2. Laymen, when given access to adequate information are quite capable of determining which goals are important to them and in which order; and

3. In dealing with professionals, laymen should have the same options normally available to them in any purchase situations.

"NADER'S RAIDERS" LOOK AT THE COMMUNITY MENTAL HEALTH CENTER

The themes of consumer participation, evaluation, accountability and professional attitudes and behavior are all played out in a depressing way in the recent critique of community mental health centers by the Center for the Study of Responsive Law (Chu and Trotter, 1972). The primary villain in that account is the National Institute of Mental Health, but lest we, who are not embroiled in the federal mental health bureaucracy, take comfort, we should also note that in this story there are portrayed no heroes, at least among the community of mental health professionals. The book paints a dismal picture of continuous failure on the part of mental health professionals to come to grips with the mental health needs of the American people. It is notable that of the twenty-six concluding recommendations which were made, at least seven were directly addressed to the topics of evaluation and accountability. It is worthwhile to examine these seven points:

(17) Centers should set up training programs for community people on the issues of human service delivery, administration of services, financing and related areas. The purpose of this training program is to prepare community people to exercise knowledgeable community involvement in all phases of the planning and functioning of the center. As much as possible, the teachers for this program should be recruited from the staff members who live in the area to be served by the center.

(18) *A governing board should be expeditiously created for each center.* The members of this board should consist of elected representatives from the community, who have undergone the training program described in the last recommendation, and a cross-section of the staff, particularly those staff members who live in the area served by the center. This single governing board should be legally responsible for receiving grants and for running the center.

(19) Each year the governing board and the staff should be required to *set specific goals* for the center and outline appropriate steps to reach these goals. This yearly statement of goals should be distributed to the community.

(20) Standards committees, with *equal professional and lay representation* should be established for every community mental health center. These committees should be responsible for promulgating explicit standards for quality in mental health care. Peer review

committees, with equal representation and functions for lay members should routinely review and grade professional performance according to these standards. Quality ratings should be made public information.

(21) Each center should be required to *establish an ongoing program of evaluation* to ascertain not only if the yearly goals specified by the center have been reached, but also all possible aspects of the center's operation (including quality of care, administrative efficiency, etc.). However, data should be collected only after the center has adequately determined why they are collecting such data, what questions will be answered by such data, and whether these questions are important. Results of this ongoing evaluation should be published yearly and distributed to the community.

(23) Site-visit teams from the regional office should include at least one community representative who is not a member of the governing board.

(24) All site-visits and *the evaluation files of the centers should be public information.*

It is quite clear from the above that according to these authors all information concerning the community mental health center, with the exception of patient files, should be available to the public. This emphasis on publicity is obviously critical to any notion of accountability. The greater the secrecy, the greater the potential for abuse. The constant emphasis on community representation, and on public dissemination of information about centers' operations, reflects the view that professionals need regulation and must be held accountable to their clientele.

CONCLUDING COMMENTS

One should understand that the issues dealt with in this paper are not easily disposed of. The underlying problem is complex and in need of continual surveillance. Simply put, this problem is one of the uses and abuses of power and privilege in a democratic society. The allocation of authority to persons and institutions empowered to act on behalf of, or against, persons is an obvious necessity in any complex societal structure. How, then, do we regulate the use of this power so that it serves the purposes for which it was intended without at the same time allowing the wielders of power to aggrandize themselves or victimize others? It is a familiar question applicable not only to military and para-military groups, or to politicians, where its application is obvious, but also to civil servants, physicians, college professors and in the specific case considered herein, the professional community associated with rendering mental health services. The general approach to all of these diverse specialists is to hold them accountable to the society which supports their work. It is when we attempt to specify to which segments of society, and to whom, and by what means, that answers become increasingly difficult to obtain.

With regard to professionals, the most reasonable approach to rendering them accountable is careful evaluation of their work, with special attention to the effectiveness of their methods in achieving consensually agreed-upon outcomes. In this connection, service populations (consumers of services) must be carefully defined, the definition plainly stated, and measurable outcomes determined. In the final analysis, it is results which count and the systematic assessment of results then becomes the basis on which a mental health service delivery system may be rendered accountable.

REFERENCES

Chu, F. D., & Trotter, S. *The Mental Health Complex, Part I: Community Mental Health Centers.* Washington, D. C: Center for the Study of Responsive Law, 1972. (Mimeo)

Cicarelli, W. "The impact of Head Start: Executive summary." *The Impact of Head Start: An Evaluation of the Effects of Head Start on Children's Cognitive and Affective Development. Vol. 1.* Bladensburg, Md.: Westinghouse Learning Corp., June, 1969, 1-11.

Evans, J. W. "Head Start: Comments on the criticisms." *Britannica Review of American Education. Vol. 1*, 1969. Chicago: Encyclopedia Britannica, Inc. 253-260.

Levy, L., & Visotsky, H. M. "The quality of urban life: An analysis from the perspective of mental health." *Urban America: Goals and Problems.* 1967, Washington, D. C., 100-112.

Miller, W. B. "The impact of a 'total community' delinquency control project." *Social Problems*, 10, 1962, 168-191.

Rossi, P. H. "Evaluating educational programs." *Readings in Evaluation Research.* New York: Russell Sage Foundation, 1971, 97-99.

Weiss, R. S., & Rein, H. "The evaluation of broad-aim programs: A cautionary case and a moral." *Annals of the American Academy of Political and Social Science*, 385, 1969. 133-142.

SECTION V

Manpower Development

Effective manpower development programs in the human services are dependent upon our having an ability, however limited, to predict the future, our having training programs and conceptual models of community life that are likely to strengthen the natural environment, and finally, a fair amount of luck. Even under the best of conditions, the problems of developing and implementing worthwhile manpower development programs are immense, diverse, and complicated.

Historically, human service delivery systems in general, and mental health delivery systems in particular, have not distinguished themselves in American life. Too often they have been rooted in the past and tied to certain ideological traditions, rather than geared to the future and the performance records of particular treatment systems. Our training models have seemed more concerned with being ideologically correct than effective, and our human service bureaucracies, more concerned with categorical order than community services.

Three major problems in human service manpower development are identified and dealt with in Section V. The first of these is the problem of overall system considerations, involving an examination of recent population trends, national and regional social policy, and training programs that hold promise for the future. These are discussed in Broskowski and Smith's chapter on "Manpower Development for Human Service Systems".

A second major problem is that of a set of operational considerations which can guide the development of training programs. In Malott's chapter on "A Behavioral-Systems Approach to the Design of Human Services", a series of behavioral principles for human service manpower development programs are discussed. In contrast to many approaches to training, one based on these principles is dependent upon empirically established results.

Finally, the problem of integrating intervention programs into the fabric of community life is dealt with in Klein's chapter on "Community Change Agents in Mental Health". The extent to which human service systems segregate themselves from, or integrate themselves into, the natural ecology of community life is a basic and far-reaching organizational decision. Some guidelines for the development of more integrative approaches are described by Klein.

Taken together, the chapters in Section V suggest some different approaches to some old problems. While it is probably unreasonable to hope that organizational decisions based on these approaches will be uniformly better than those of the past, it is the case that it would be difficult to do much worse.

Anthony Broskowski

Terence P. Smith

If human service systems are to develop effective programs, they must have the capacity to respond to a changing environment. Only recently have these systems begun to develop that capacity. Our historically limited approaches to human service problems, reflecting both the legislative and the human service bureaucracy's categorical views of the world, have begun to give way to more flexible approaches to solving human problems.

In the following paper, Broskowski and Smith[1] *discuss many of these changes, including alterations in population, evolving social policy, and new conceptual approaches to the planning and administration of human services. The impact of these changes, to the extent that these might be predicted, are assessed, and the authors submit a number of short and medium-range predictions for the human services. While they may or may not be entirely accurate, we ignore considerations of these broad trends at considerable peril to both our human service systems and, more importantly, the public. At a minimum they are essential concerns for the development of viable manpower policies in the human services.*

CHAPTER 14

Manpower Development for Human Service Systems

In 1803, the British created a civil service job calling for a man to stand on the Cliffs of Dover with a spyglass. He was supposed to ring a bell if he saw Napoleon coming. The job was abolished in 1945.

Increasing complexity and rapid change have become *clichés* in the discussion of contemporary issues. Nevertheless, these two factors emerge as the final background in an analysis of manpower developments for human service systems. These two factors make rational analysis and planning for human services manpower even more critical and at the same time more difficult.

To understand human services manpower development it becomes necessary to analyze the complexity and change taking place in the much larger environment in which manpower systems operate and adjust. For purposes of this paper we have arbitrarily differentiated this total environment into an analysis and review of trends in population, social policy, and human services program delivery. Population, policy, and programs are each interrelated and singularly, or in combination, will certainly influence the future human services manpower development, including both factors of supply and demand. This presentation does not attempt to predict with precision the future course of human services manpower since complexity and change make it impossible to do anything but dimly perceive the most probable short range developments.

POPULATION TRENDS

The 1973 *Manpower Report of the President* (Department of Labor, 1973) stated:

> By far the most important demographic development of the past decade has been the clearly marked transition from the three-child to the two-child family average.

For several reasons the trend toward the two-child family may be expected to have a profound influence on the manpower picture of the next decade. There can be expected an increase in the number of working wives, many of whom will be returning to the labor force and seeking jobs in the human services sector. There will be an increased demand for certain types of human services that have a relationship to these new family life styles. The average age of the work force will increase with a gradual decline in the proportion of younger workers. This age pattern among workers will also apply to the human services industry and further influence the mix of personnel and training.

Increasing numbers of women in the labor force will have several significant effects on human services manpower. Participation in the labor force by women now tends to follow a bimodal distribution according to age. The figures for 1971 show that between ages 20 to 24 there was a peak participation close to 60 percent of the female work-age population. Between the ages of 25 to 34 there was a decline in participation. After age 34 there is an increase in participation, reaching a secondary peak during ages 45 to 54 and falling off abruptly after age 64. This bimodal pattern is accompanied by a marked tendency for older women to re-enter the labor force in jobs which are different from the jobs which they had initially. (Similar career changes are becoming increasingly frequent for men as well. The growth of continuing education and retraining attests to these conditions.) Increasing militancy among women for equal pay and equal job opportunities will require major shifts in present manpower programs and human services educational policies. Significant efforts will have to be expended to meet the demand for relevant career training by women seeking human services employment after childbirth years.

The shift to the two-child norm will have further implication for human services by virtue of the demand it will create for particular types of services to particular age sectors of the population. Given lower fertility rates, the proportion of dependents under 16 years old will decline steadily from 52 percent to 46 percent by the year 2000; while the proportion of aged dependents will rise from 14 percent in 1970 to 18 percent by the year 2000. Demands for some services associated with child-rearing and education may be reduced, particularly those not directly related to child care of the very young. Demand for and availability of day-care services will, in turn, increase the number of women in the labor market. Demands for human services for the elderly will increase. Manpower training programs must consider such probable shifts in service supply and demand. A few examples may help to illustrate some of the relationships between population changes and service supply and demand factors.

There will be great swings in the *relative* sizes of the elementary, secondary, and college populations for the remainder of this century. Today's surplus of elementary teachers may be tomorrow's undersupply of teachers for colleges or special vocational schools. The increasing mean and modal age of the population will create shifts in the *type* of medical care being demanded. While some predictions point to an increasing need for *all types* of medical personnel (Department of Labor, 1973), the H.E.W. Assistant Secretary of Health, in a speech to the Association of American Medical Colleges, suggested the possibility of a physician surplus due to declining population growth rates and new ways to increase the actual productivity of practicing physicians (Edwards, 1973). Such trends in service supply and demand, whether in education or health, can be selectively enhanced or offset by changes in educational or health service technology. New technologies, in turn, replace the need for some personnel while requiring that others be available. The complexities increase *ad infinitum.*

In summarizing the far reaching implications of population changes on manpower policy, the 1973 *Manpower Report of the President* (Department of Labor, 1973) stated:

> Finally, some attention must be paid to the long-term implications of possible changes in the life styles of the American people. If slower population growth implies a continued growth in material affluence, it also implies the opening of a greater range of options concerning work, leisure, job retraining, and continued education. The nature of the manpower policies which emerge during the remaining years of this century will necessarily depend upon the responses developed to meet this new range of options.

FEDERAL SOCIAL POLICY

Sufficient time has elapsed since President Nixon's first election to detect a clear and major switch in federal social policy regarding the economics and organizational pattern for human services. Furthermore, many of these existing trends are likely to continue regardless of any future changes in the executive branches of the federal government. The present policy pattern is likely to persist because it derives from a complex set of factors not totally under the control of the executive branch of government nor necessarily based on a particular political ideology.

The Income Strategy

One major policy change has been a reduction in the support of government-sponsored service delivery, characteristic of the The Great Society, and an increase in the income-to-citizens strategy, characteristic of the New Federalism (Nathan, 1973).

On January 1, 1974, the Social Security Administration began operating the Supplemental Security Income (SSI) program, a cash payment to the aged, blind, and disabled that is designed to remove all recipients from below the poverty-income line. Persons who will be receiving SSI presently constitute one-half of the caseloads of state and county welfare departments and it is expected that the caseload once under federal auspices will double from three million to six million (Nathan, 1973). Supplemental Security Income, along with raising Social Security payments, Medicaid and Medicare benefits, food stamps, other in-kind benefits (e.g., health maintenance organization premium payments, subsidized housing), and reduced income taxes for the poor, will have increased the cash payments and in-kind support programs of the federal government to 40 percent of the proposed Fiscal Year 1974 Budget.

Additional plans are being made to further reform the welfare system through a program of work bonuses and income support tied closely to the available job market. Within five years we will most likely see a National Health Insurance Program. Even the educational program of student loans in preference to large training grants to institutions reflects the policy of "cashing-out" the institution-based service programs of The Great Society for the increasing-personal-income strategy of New Federalism.

The preference inherent in the personal income policy is to give to the citizens, as directly as possible and consistent with need, the cash benefits of taxes, rather than having the government pay for expensive services, of which only a small dollar value may trickle down through the levels of government bureaucracy and service providers to those who may need such services. In essence, this policy shifts the purchasing power to the individual citizen and allows a market-place economy to determine the mix of human services through laws of effective demand and supply. The income policy will likely stimulate an increased demand for some types of human services, such as medical care, while it inhibits the development of such services as mental health consultation and education. In general, direct services to families and the individuals may flourish while indirect services, particularly those with no demonstrable and immediate effects, may languish (Broskowski and Baker, 1972). It is still not clear which human services will be funded by recipient payments, third-party payments, or government grant support.

Revenue-Sharing Strategy

Another major development in federal policy is the reality of general revenue-sharing. Over 40,000 state and local government jurisdictions are presently receiving payments which are not specifically restricted to rigid categories of expenditure. Local and state governments are only minimally restricted in how they may use these funds for whatever priorities the local constituencies determine. The leveling-off and reduction of categorical federal funding will further erode the highly refined categorical service delivery strategy and require a re-examination of human services at the local level. Erosion of categorical service funding will erode categorical training as well. We are also likely to see the enactment of special revenue-sharing legislation in broad categorical areas.

As the citizens and local and state governments gain increasing control over the flow of funds, they will have increasing influence over manpower training, allocation, and distribution decisions. These new decision-makers may or may not improve upon the decisions formerly made in Washington and influenced by categorical constituencies, professional associations, and other special interest groups. While many influential groups will undoubtedly remain influential, the base will have broadened and the level of decision-making will be brought closer to the community.

The reduction of categorical services funding and delivery will also promote greater experimentation with the major models of public administration at the state and local level which, in turn, will influence local manpower systems. In brief, there are four major models by which public services can be administered.

1. Government owning and operating services.
2. Government regulating private utilities or service monopolies.
3. Government contracting for services.
4. Citizens purchasing services with cash or vouchers supplied by government.

Each model has its advantages and disadvantages and each is governed by somewhat different principles of economic supply and demand. Depending upon the degree of standardization and coordination involved or required in the delivery of services, one model is to be preferred to another. For example, standardized electrical and water services to the public appear best administered by either government-owned-and-operated models or government-regulated utility models; whereas, the uniqueness-factor inherent in medical care or psychotherapy, would suggest that these services would best be administered through contract models or voucher systems.

Medicare and medicaid are essentially citizen-voucher systems. Education is now experimenting with the voucher model as opposed to the traditional government-operated school system approach. Vocational rehabilitation and sheltered workshops have usually been services for which the government contracts. In contrast, the greatest government dollar expenditure for mental health services has been in large state-owned-and-operated mental hospitals. This latter pattern may not be most appropriate, but developed out of complex historical reasons peculiar to mental health. Even some correctional programs are now being contracted-out in preference to the state's owning its own adult or juvenile "correction" facilities.

State-owned-and-operated services are generally less flexible and responsive to needed changes based on new technologies or treatment approaches. The resistance to change is over-determined and includes such factors as government civil-service systems that must guarantee job-tenure, patronage among employees, and strong politically-based constituencies that have vested interests in the *status quo*. While the contract model provides flexibility by passing off to the contractor the expenses and risks of required change and adaptation, the short-range costs may be greater due to the addition of a profit margin and the larger salaries paid in the private sector to compensate for government job-tenure security.

The critical questions of which model has the best cost/benefit ratio for which types of public services, over which time periods, and under what different types of circumstances, has yet to be answered. It may never be answered! Under most circumstances we are likely to observe a mixed set of models operating simultaneously or in coordination. What is certain is that more experimentation with alternative models and mixed models will be taking place. Our comfortable patterns of service delivery will be disrupted, and this disruption, in turn, will have profound impact on manpower training and development.

TRENDS IN HUMAN SERVICES PROGRAMMING

Other chapters in this volume have already explored some of the major programmatic trends for human services. We would like to briefly review those which we think will have the greatest impact on human services manpower systems. While some of these changes could have been equally subsumed in the previous section as changes in federal policy, which indeed they reflect, we have arbitrarily decided to cover the discussion of program changes in a separate section.

Services Integration and Organizational Management

The major trend in the delivery of human services has been the recent emphasis on the integration of existing resources. Rather than build new programs and agencies as in the 1960's, most policy makers have agreed that limited resources require a strategy of increasing service productivity and efficiency through the effective integration of current programs. An increasing number of analyses point to the need to eliminate the wasteful and counterproductive duplication and fragmentation among existing resources. For example, since 1930 the Congress has legislated several hundred separate health, education, and social welfare programs. Most of these programs are narrowly categorical and unrelated to one another. The general complexity, rigidity, red tape, inefficiency, and ineffectiveness of these programs have become common knowledge.

The thrust of the integrated services strategy has been to change the *total delivery system* so as to increase its accessibility, comprehensiveness, continuity and accountability. Evaluation data on model services integration projects have indicated a number of factors that can either facilitate or inhibit the multiple types of linkages that are required to integrate existing delivery systems at the local level (DHEW, Social and Rehabilitation Service, 1972). The attitudes and training of service personnel are among those factors that are critical to the achievement of an integrated service system.

In an increasing number of states, integrated systems are being developed to plan, deliver, and finance social services, using estimates and evaluations based on measurable units. Although the federal government never mandated the H.E.W.-designed system of Goal Oriented Social Services (GOSS), the related national GOSS-training program of 1973 has caused many state human service agencies to re-examine their service systems and to adopt some or all of the goal-oriented approach with the sub-systems of: Management Information, Manpower and Training, Cost Analysis, and Programming and Financial Planning. Accountability is built in with the determination of specific operational targets toward which all efforts and activities are directed, and against which effectiveness is evaluated. The focus is thereby shifted to the outcome of the services — the product and not the process. Specific examples of such statewide systems and their influence on manpower developments are presented in later sections of this chapter.

Integrated service networks will demand a greater degree of interdisciplinary work, including the full utilization of professions that were previously not considered as critical to the human services system. Architects, engineers, land planners, lawyers, economists, and information specialists are just a few of the many professions with useful contributions. New treatment technologies, new joint service locations, and new administrative and finance models will create problems requiring the attention of many professions. No one profession can be expected to dominate.

Major intra and interorganizational changes include the greater use of new management and planning tools, including such devices as management and client information systems, cost-accounting systems, shared-management services, and complex organizational network structures such as multi-service centers and service consortiums. Due to the high costs of delivering human services (Gruber, 1973) greater emphasis is being placed on management and service practices that will increase productivity and efficiency, criteria which many human service professionals have been inclined to disregard when working in government-sponsored service systems. In general, we can expect a greater concern for the economics of human services delivery (Rosenthal, 1974) and manpower development (Klarman, 1969), including a consideration of benefit/cost ratios. For example, one econometric analysis for the evaluation of manpower programs points out the frequently forgotten fact that one of the greatest costs of a training program is the foregone earnings lost to the participants during the program and post-program period (Rostker, 1973). In general, the opportunity costs of educational programs are the largest single cost item.[2] One analyst has estimated that the foregone wages of students while in high school account for about 75 percent of their total educational costs (Schultz, 1963).

De-institutionalization and De-professionalization

Another common program trend is the major shift toward the use of community-based programs and a reduction in the use of large institutions. The movement toward institutional reform is evident in the shutdown of state mental hospitals, decreasing reliance on nursing homes for the aged, and a turning away from large correctional facilities. Those uses of institutions that do continue will be more specific and specialized in their goal-orientation, and considered to be only a part of an integrated care system and individual treatment program. Institutionalized service programs will become smaller, more refined, and more differentiated.

One of the most noteworthy shifts in program delivery systems has been the rapid emergence of the self-help philosophy of treatment (Caplan, 1973). More and more people, many of whom were formerly ignored by professionals, or socially stigmatized, are organizing their own groups to help one another. Alcoholics, drug-addicts, widows, skid-row bums, child-beating parents, former mental patients, and handicapped persons of all kinds, are finding that help can be faster, cheaper, and more lasting, without necessarily involving professional care.

The self-help movement is really only part of a larger trend toward using naturally existing support systems for preventative (Broskowski and Baker, 1972) and ameliorative purposes (Caplan, 1973). The emphasis on support systems will recast the roles of professionals and paraprofessionals in the service delivery mix. Trained specialists in particular human services (e.g., mental health, rehabilitation, welfare, education, etc.) will be asked to contribute to these new groups and support systems by helping to initiate or organize new ones. In addition, professionals can offer consultation and education to the organizers and members of these support systems (Caplan, 1973). Some self-help groups are based on an ideology that is oppositional to professionalized service and it can be expected that such groups will be exempt from professional assistance.

A related trend is the increased use of volunteers as part of organized delivery systems. The chapter in the present volume by Harold Demone spells out some of these developments. Federal regulations and federal support are providing further accelerations to volunteerism.[3]

The push for integrated and community-based services, coupled with the trends in self-help, volunteerism, and natural support systems, will have profound impacts on the professional manpower pattern. The role of professionals as direct service providers may become more constrained as people find that they can get help from each other and spend their discretionary income on other types of goods or services. These program trends will also intensify the generalist-specialist dilemma. Lower level personnel will increasingly be trained as specialists in highly technical tasks but they will also seek ways to further generalize their experiences as an adaptation to obsolescence and changing demands. Higher level personnel will need sufficient expertise to train and supervise the complex mix of lower level personnel. Furthermore, the traditional professions must find new roles for themselves as some of their own activities are taken on by persons with task-specific training. Professionals must seek and solve new challenges to justify their status, authority, and salary levels. As more treatment activities are carried out by personnel with lower salaries and less training, higher level personnel must assume more demanding roles in policy formulation, planning and administration (Demone and Harshbarger, 1973), research and development (Broskowski, 1971), and program evaluation (Suchman, 1967; Weiss, 1972). Such changes are now appearing in the formal curricula and field placements of professional training programs.

Other important program trends include the increasing array of treatment methods being used in the human services field. Some of these newer methods can be classified as types of psychotechnology (Lanyon, 1971), including the use of drugs, new testing and screening methods, and behavior modification techniques. More and more frequently the interventions are being focused within the specific environments in which the problems occur, such as homes, schools, and work settings, rather than in the artificial confines of an agency or private office. There is an increasing interest in the family as the basic unit of service delivery. Record systems, organizational arrangements within social service agencies, and methods of treatment are emphasizing the family as a critical system for integrating disparate services and developing continuity over time.

HUMAN SERVICES MANPOWER DEVELOPMENT

Given the trends in population, federal domestic policy, and human services programming, it is not too difficult to speculate about a wide array of manpower issues for human service systems. Each of the trends that have been reviewed necessarily interact with one another, sometimes reinforcing one another, sometimes cancelling each other out. The influence of these changes on human services manpower systems will undoubtedly create its own feedback which in turn will have a reciprocal effect on population, policy, and program delivery. The following analyses are necessarily limited to only a few major issues.

Estimates of Supply and Demand

A review of the national supply and demand projections for each of the specific categories of human services personnel will quickly indicate the difficulties of making any generalizations about national trends. To begin with, there is no systematic classification system that allows one to aggregate the specific occupations that belong to the human services industry. Put another way, there is not yet a clear definition or consensus on the boundaries of the human services system that would allow one to count who's in it and who's out of it. There are presently some attempts at the level of state government to develop general human service personnel classification systems (Marsh, 1973), and to further project statewide estimates of supply and demand. These efforts will be more fully illustrated later in this chapter with examples from Maine and Utah. Although the U.S. Bureau of Labor Statistics does publish national projections based on *present* job descriptions, their classification system will have only a partial overlap with the probable job classification systems of the *future*. Like our present funding and delivery system, job classifications are presently based on categorical systems often centered around single professions or related professions (e.g., health professions). The fragmentation of manpower categories is paralleled by a fragmentation of manpower planning efforts (David, 1969).

Given a coherent classification system, all supply and demand projections must still make a host of assumptions about the nation's future economic and political status. The difficulties were well stated by the Department of Labor, Bureau of Labor Statistics (1970), in a publication on the supply and demand for college graduates:

> Estimates of future supply also are greatly influenced by their underlying assumptions. For example, wage differentials, social status of occupations, the availability of training, the nature and extent of student financial support, the length of training, and immigration laws all affect the supply of workers in a particular occupation.

In making such assumptions, many manpower analyses simply assume that present trends in college interests, entry patterns, subject matter appeal, salary level at graduation, etc., will continue as presently constituted. Yet we know that these factors may shift considerably over the years.

For the above reasons, it is not feasible at this time to present any neat packages of quantitative projections on a national basis. Future refinements in computer simulation of the multiply-interacting supply and demand functions, together with a refined human services personnel classification system at the state and federal level, will allow for some short-range predictions to be made. Readers interested in specific manpower projections for the mental health field are referred to Arnoff, Rubinstein, and Speisman (1969), and an excellent summary review by Williams (1972).

Human Services Manpower Training and Educational Options

Manpower training programs must change along a number of dimensions to adapt to the above changes in population, policy, and programs. Training programs must be designed for diversity, flexibility, and a greater degree of integration with the multiple human *service systems.* Training programs must move out of the closed classrooms and into a wider number of open field settings. Based on these needs we can expect to see more training being totally sponsored by human service agencies. Using apprenticeship and work-study models, agencies will increasingly recruit and train for their own manpower needs, independent of academic auspices or through the purchase of limited academic resources. Continuing education will become a more predominant trend, fed by the need to keep up with the rapid changes in service technology and service demands. In-service training and on-the-job training units will become regular organizational features of large-scale planning, research, and service delivery systems.

An H.E.W.-commissioned study, "National Policy and Higher Education," recommends that the federal government use its funding policies in higher education to foster greater flexibility, autonomy, and diversity among institutions of higher education (Department of H.E.W., 1974). The report spells out the potential alternatives to the present rigid and lockstepped four-year system that undermines competency, promotes credentials, and stimulates the proliferation of professional guilds and accrediting agencies. Indicative of the trends discussed earlier in this chapter, the report recommends that grants go to students rather than institutions. The report also suggests a type of "G. I. Bill for Community Service," allowing students to accrue credits for periods of work in national or regional service programs. In brief, "the major thrust of the report is the need to loosen up the system so it can become more responsive to the demands of the student market" (Holden, 1973).

To the extent that training programs will be less insulated and buffered from changes inherent in a market economy, they should be designed to accommodate abrupt shifts in size. Strategies for accommodation are likely to be reflected in a growth of such institutional policies as the elimination of tenure, which in turn will be counteracted by faculty demands for larger salaries or unionization to offset job insecurities. Since students will find programs to be less stable in size and duration, they will increase their demands for short-range relevancy and competency-based training. The increase in options for content and method will also stimulate more short-range shifts in personal career development.

The volume of training will broaden at the academic entry levels, with increasing growth for community colleges and the number of associate arts degrees in human services. All types of new career training programs will continue to increase (McPheeters, 1969; Young and True, 1973) regardless of the multiple barriers to adequate job status and career development that presently exist (Vidaver, 1973). A good sampling of the multiple levels of innovative mental health training programs, both academic and non-academic, for the roles of service, teaching, research, and prevention, are provided by N.I.M.H. (1971).

The concept of career ladders is presently being expanded to a concept of career "lattices" (Vidaver, 1973). All levels of personnel will increasingly change career interests as they become exposed to, and move about in, an increasingly integrated service system. New careers and second careers are already a familiar occurrence among middle and upper-income businessmen and women returning to the job market.

The increased diversity in the content and methods of human services training will stimulate the creation of an even greater number of new and complex career options, as formerly independent fields of study are linked through new policies and integrated service delivery systems. The greater number of options available to potential students or trainees will also stimulate several reactions among existing educational and service systems. Diversity of choice will increase educational costs due to the higher investments needed to carry out educational planning, marketing, and related maintenance functions. The savings of simplification, consolidation, or monopolization will be lost. The competition for potential trainees will increase. Increased diversity will also heighten concerns within traditional service and educational systems for regulation and standardization of credentials. Well-accepted professions will resist the development of new professions. If resistance fails, the entrenched professions are likely to try cooptation or direct control of new professionals. Established training programs will demand that newer roles and practices be evaluated while resisting any attempts to examine their own outcomes.

The evaluation of training programs will be a major concern of the future. The types of evaluation will depend on the mechanisms used to finance the training and provide incentives for better training performance. Programs sponsored by direct government support will be required to demonstrate their effectiveness and efficiency through formal evaluation designs (Broskowski and Khajavi, 1973; Broskowski, 1973). Programs that are funded by trainee-controlled vouchers, stipends, loans, or tuition, are likely to rise or fall based on their value as determined by the consumers (i.e., the trainees).

Letting the trainee make the choices will only partly solve some of the problems of diversity, costs, and evaluation. Our existing large-scale commercial and industrial systems are good illustrations of how consumers are *not* effective in creating organizational responsiveness because large systems can effectively control their own environments (Galbraith, 1973). There is much that can and will be said for educational models that rest upon social contracts that buffer educational systems from a totally free economic market.

Prior to World War II, professional training programs were not highly subsidized by direct grants from government. Large grants to professional education programs were developed out of a policy to stimulate rapid increases in selected professions. Market choice mechanisms were considered either too slow to respond to immediate needs or too easily influenced by short-range personal choice variables to respond to national priorities.

Personal market choices can be expected to operate effectively only when the potential consumers of education are quickly and sufficiently well-informed as to the long-range costs and benefits of their early choices. Since long-range predictions become increasingly unreliable, multiple branching points for secondary decisions must be made available. Since all the costs and benefits of various human service careers can be only dimly perceived at the time that one is forced to begin making choices, the value of personal planning breaks down and the need for such systemic options as continuing education or career change are increased. Furthermore, the research on career choice indicates that few persons make early career choices using eventual financial payoff as a major criteria (Zytowski, 1968). Instead, youthful decisions are more frequently influenced by such chance factors as the availability of early role models, parental values, and short-range adolescent needs and expectations. Career planning based on facts and knowledge of available alternatives is seldom the case. We also know that youthful naiveté and altruism erodes with age and rising inflation. For such reasons the system must provide multiple options for career development.

A Systemic View of Manpower and Training

A system might be defined as a way to organize our knowledge and resources in order to achieve some goal within a specified time frame. Part of any large service system is the manpower sub-system. Part of the manpower sub-system is the training sub-system. There is a definite trend in current management theory to consider manpower and training functions as integral sub-systems within the total organization (Gross, 1964). Manpower is usually an organization's most costly and valuable resource. It is a critical factor if the organization is to achieve its goals. In human services manpower is doubly important for it is the raw material as well as the product.

A service system's goals may be clear, but sub-objectives are frequently unrelated or conflictual. In the process of analyzing and defining the activities necessary for the attainment of an agency's service goals, many personal values, prejudices, and needs emerge. Frequently, these personal goals are at odds with those of the agency and are sometimes at odds with each other. Although such conflicts are rarely faced or resolved, the dynamics of organizational development and effectiveness require that personal and organizational goals become merged (Likert, 1967). It is one of the major roles of training sub-systems to bring about such a merger. Training units, however, are frequently without their own clearly defined goals and are usually required to function in areas considered peripheral to the central organizational issues. Training units have also been called upon to perform in crisis situations that do not allow sufficient time for the planning of training tasks. Because of their key role for the total organization and the large investment in manpower, training sub-systems are increasingly being forced to develop a goal-oriented approach (Tracey, 1968). Plans, budgets, and measurable goal criteria are rapidly replacing hip-pocket and crisis-oriented training activities.

The organizational role of the trainer is also evolving from the position of a content instructor to one of administrator and organizational development specialist (Nadler, 1970). More generally, training sub-systems have been expended to include personnel functions, and raised to a position of planning and policy-making within the organization (Argyris, 1964).

Many of the elements which affect a manpower and training sub-system are complex and must of necessity remain uncertain. On the other hand, there are definite elements which can be identified that will enable us to approach manpower training and development with more reverence and intelligence. For example, there are a finite number of tasks or services which any agency can deliver. These tasks can be identified, catalogued, and combined in a large but finite number of ways. Usually such service tasks have been grouped into job descriptions on the basis of tradition and professional guild authority. Upon examination, many such jobs reveal areas of boredom, repetition, and waste. It is conceivable, however, to computerize the total number of tasks so that the goals of consumers, agencies, and personnel can be integrated and optimized. Although such systemization may seem staggering, a great many components are already in place. Examples of such efforts are illustrated by changes currently taking place in statewide service delivery systems.

Statewide Systemic Approaches to Service Delivery and Training

In addition to the changes taking place in formal academic training programs, manpower development will be directly influenced by changes occurring in the actual service delivery systems at the state and local level. As noted earlier, the Department of Health, Education, and Welfare has exerted pressure on state-operated human service agencies to integrate services at the local community level, establish goal-oriented measurable services, and separate money payments from social services. These pressures were instigated because of ever-increasing costs and a general lack of confidence in public service agencies. In an effort to respond to these pressures many states have developed new approaches. Maine, Pennsylvania, Florida, Michigan, and Utah are just a few of the states where policy changes have been made and efforts expended to develop a systems approach to service delivery.

Maine, for example, designed its human service system around units of service delivered by staff specialists in counseling, protective service, housing, and many others. The State makes extensive use of purchased services from the private sector. The client is seen initially by a "programmer" who makes an assessment with the client and sets client-appropriate objectives. A contract is made with the client to deliver specific units of service in a definite period of time. The service delivery specialists report on the units actually delivered through a statewide management information system. The service cost is the product of the number of units times the per-unit costs.

Statewide planning and goal-setting mandates that some services be made available while others are determined by each Regional Office. Both mandated and optional services are determined through a process called Target Objective Mix. The regional director surveys the community, and using a variety of indices, determines the size of the target population, sets objectives for that population, determines what mix of services are needed to meet the objectives, and estimates the costs of delivering the services by unit measures. The services are thereby specified, measured in terms of time and cost, and inventoried to be drawn upon as required by client needs. The service-needs assessment includes a staff survey of the current caseload and inputs from regional and state advisory committees. These methods identify presently-serviced as well as most unmet needs. Decision-making involves the consumers, program staff, management staff, a central office planning and coordinating committee, as well as the state government planning, budget and legislative units.

The installation of this new service system has required a massive statewide training effort to explain the new policies, teach new tasks and skills, and develop different staff attitudes. Because the use of many of the services by citizens is voluntary, the service personnel have to be given the skills to further train the potential clients to use the new approach.

As new service tasks are identified, new positions are created that do not necessarily require a professional degree. Work experience and formal credentials have equal validity as criteria for position and promotion. The skills of planning, management, and evaluation have been taught to existing staff or recruited from outside the system. Maine is presently developing a manpower planning capability through a contract with the University of Maine. This planning capability will hopefully allow the State to forecast manpower needs and then design and support the specific training programs to meet these needs (Davis, 1973).

Probably the most significant change for manpower development has been the enhanced and integrated position of manpower policy in the overall organizational structure. Job description, classification, and restructuring, and personnel selection, promotion, and training have become integral management tasks within the context of services delivery.

Functional Job Analysis

A good example of a systemic statewide approach to manpower training is provided by the efforts of the State of Utah's Division of Family Services in pioneering a unique and powerful technology for manpower planning called Functional Job Analysis (Lewis, et al., 1972). Functional Job Analysis (FJA) provides a model for linking staff development and personnel considerations with program service goals, and a way of developing the basic work information essential for job design, job restructuring, job enrichment, career planning, selection, evaluation, promotion, training, and manpower planning. In short, FJA is a logical systems approach to meet the goals of the agency *and* the employee.

The FJA technique has been refined by Sidney Fine of the Upjohn Institute under a grant from the Department of Health, Education, and Welfare (Fine and Wiley, 1971). Over six hundred tasks have been assembled into a data bank. These tasks are the results of contributions from many states and constitute a bank of tasks common to social service agencies. Similar efforts have taken place in rehabilitation agencies. With a minimum investment of time, a given agency can review the various tasks and adapt them to their own particular service programs. Various tasks can be combined in a variety of ways to create totally new and unique jobs.

Each unit of work or task is analyzed according to various dimensions and summarized on a single card. An example is given in Table 1. Starting at the top left side, the task is analyzed according to its degree of difficulty in the areas of data, people, and things, and rated on a scale of 1 to 6. The task is next broken down into the percentage of time spent in each of these areas. The next dimension is a Worker Instruction Scale which determines what amount of direction or discretion (1 — 8) the task requires. The next three dimensions are adopted from the "General Educational Development" index of the Dictionary of Occupational Titles (1965). These dimensions define the level of reasoning, mathematical, and language ability needed to perform the task. The last box is a code number for indexing the task. When using FJA, the sub-goals of the agency's sub-systems must be clearly stated and related to the mission and goals of the larger system. The task statement is a precise statement of an element of work necessary to accomplish the objective. The task is then broken into a) performance standards, and b) training content. The former serves as a clear standard against which the worker is to perform. The latter is a breakdown of the specific training elements that are needed for adequate performance.

These task statements provide a common denominator to analyze the service demands, the goals of the agency, and those of the individual worker. Budget estimates, training plans, recruitment, selection, promotion, career development, upward mobility, and an ability to change as the needs of clients shift, are some of the obvious consequences of such a system.

One important element in the FJA approach is that a worker is not prejudged and locked into a global caste system based on the presence or absence of educational credentials. Each worker is evaluated on the ability to perform a set of tasks in a manner which meets clear criteria. The skills necessary to accomplish these tasks can come via a college education, life experience, in-service training, or continuing education.

Modular Education

Another systemic approach to manpower training and development is illustrated by the growth in modular education, which can bring together the service delivery system with the established education system

TABLE I

A Sample Task Description Card From an FJA Task Bank

Data	People	Things	Data	People	Things		Reasoning	Math	Language	
W.F. — Level			W.F. — Orientation			INSTR.	G.E.D.			TASK NO.
3B	3A	1A	35%	60%	5%	3	3	1	4	W.E.6

GOAL: To place clients in appropriate jobs through placement counseling.

OBJECTIVE: 80% of former clients gainfully employed after 1-year follow-up.

TASK: Suggests/Explains to client reasons for making favorable appearance, and specific areas in which he needs improvement to conform to local standards/expectations, in order that the client makes job applications appropriately dressed and groomed.

PERFORMANCE STANDARDS

Descriptive:

- Information given client is complete, accurate, and clear.
- Appropriate approach/manner/attitude.

Numerical:

- In less than X% of cases, prospective employers report inappropriate dress and grooming as a result of inaccurate, incomplete, or unclear information provided by worker.
- Less than X% of employers complain of worker's manner.

TRAINING CONTENT

Functional:

- How to explain grooming standards to client applying for job.
- How to persuade client to meet appropriate grooming and dress standards.

Specific:

- Knowledge of client's situation.
- Knowledge of local grooming and dress standards.

in a joint effort characterized by flexibility and responsiveness. In modular education, specific tasks are organized into larger units or modules with the element of time being a major determinant of modular units (Putnam and Chismore, 1970).

At the Human Resource Institute of the University of Maine several hundred such modules have been developed in close collaboration with the State agencies for health, mental health, welfare, rehabilitation, aging, and corrections. The outcome is a training program for human service occupations ranging from the level of an associate arts degree to beyond the master's degree. The general categories covered by the training include counseling, community development, and human services management. Specialization into a particular discipline within any one category is possible at any level. Any one or a combination of modules can be built into an agency's in-service training program to meet the particular needs of the individual learner as well as the agency. Each module is designed on the basis of a functional task analysis at the service delivery level. The modular education plan is further integrated into the State's broader plans for service delivery (Davis, 1973).

As in Utah, this systemic approach has several advantages. Both agency needs and employee needs can be considered. The learner is not locked into a narrow track nor homogenized into a large academic environment. The credentials monopoly is short-circuited by allowing entry and exit at any point in the educational continuum, including university-based or agency-based learning experiences. To achieve individualized curriculum and self-paced learning, a full range of educational experiences and technology are brought to bear upon the training, including the extensive use of computers and audio-visual equipment. Finally, training and education are more directly influenced by the feedback from service delivery systems.

Trends in the Philosophy of Manpower Training

A manpower trend worthy of special note is the increased attention being given to the special educational methods and procedures that are uniquely appropriate to the *adult* learning process. Andragogy is the word used to describe the philosophy of how adults learn in distinction to pedagogy or child learning (Knowles, 1970).

Andragogy is based on the assumption that the adult wants to be able to do a job well, wants to obtain the knowledge for this purpose, wants to share with others the life experiences related to this goal, and can learn best in a non-threatening environment. The distinctive assumptions of andragogy and pedagogy are summarized in Table 2. The educational profession has been paying more and more attention to these assumptions as programs in adult and community education have grown to meet the development needs of a society with rapid job changes, second careers, and leisure interests.

TABLE 2

Differences Between the Teaching of Children (Pedagogy) and the Teaching of Adults (Andragogy),
Adapted from Malcolm Knowles (1970)

I. Major Assumptions	Pedagogy	Andragogy
1. Teacher-learner relationship	learner is dependent	learner is self-directed
2. Past experience	of little worth	learners are a rich source for content
3. Readiness	biological development or social pressures	developmental skills of career or social roles
4. Time perspective	postponed application	immediate application
5. Learning orientation	subject-centered	problem-centered
II. Process Elements	**Pedagogy**	**Andragogy**
1. Climate	authority oriented, formal, competitive	collaborative, informal, mutual respect
2. Planning	by teacher	mutual planning mechanisms
3. Diagnosis of needs	by teacher	self-diagnosis
4. Formulation of objectives	by teacher	mutual negotiations
5. Design	by logic of subject matter and content units	sequenced in terms of readiness and problem units
6. Activities	one-way transmittal of charisma	experimental and interactive mechanisms
7. Evaluation	by teacher	re-diagnosis and mutual evaluation

A systemic approach to manpower training that uses the principles of andragogy must develop structures and processes that conform to the following criteria (Ingalls, 1972):

1. A climate that satisfies basic human needs and is free from distractions which may cause physical or psychological discomforts.
2. A structure for mutual planning among learners and designated educators.
3. A process for determining the multiple needs, interests, and values of the learners and the institution or agency, and, insofar as is possible, a determination of their relative priorities.
4. A process for mutually setting the learning objectives, based on the tasks that need to be accomplished.
5. A process for shared responsibility in the design and implementation of the learning program.
6. A process for the shared evaluation of the learning design and the assessment of additional learning needs to be fed back at step three.

SUMMARY

In this chapter we have tried to project some of the major future trends in human services manpower development against broader changes in population growth, federal domestic policy, and service delivery programs.

We see the future as one of increased diversification in manpower options and a range of models that go beyond the tradition of specialized and credentialed training in formal academic settings. New human services careers at all levels will emerge as program changes promote a greater integration of policy, planning, and service delivery. Training will be more closely integrated with service delivery and less buffered from the exigencies of a marketplace economy. As with service programs, training will be increasingly subject to formalized evaluation, including considerations of costs and benefits. An understanding of the multiple forces that influence manpower sub-systems requires a total perspective of the human services "system" that transcends the limited views of what the trainers may like to teach or what is currently fundable by the federal government. The manpower and training decisions of the future will be more influenced by the consumers of training, including the learners and the service delivery sub-systems.

Our analysis has generally ignored the specific content of future training efforts. At the moment, the greatest gaps appear to be in the opportunities for training in human services management. In general, the content decisions will become less critical as greater options develop for in-service training and continuing education.

While our analysis has not lead to any precise forecasts, we hope that it will provide some conceptual comfort to those who are concerned about manpower training. While the future of the human services training system portends turbulent change, our understanding of its larger environment can help us to negotiate these changes and develop a better system.

FOOTNOTES

[1] This chapter was co-authored by Terence Smith in his private capacity. No official support or endorsement by H.E.W. is intended or should be inferred.

[2] Opportunity costs of a program are considered by economists to be equal to the value of the benefits that are expected from the foregone alternative.

[3] For complete information on volunteerism, the reader can contact The Clearinghouse, National Center for Voluntary Action, 1735 I Street, N.W., Washington, D. C. 20006.

REFERENCES

Argyris, C. *Integrating the Individual and the Organization.* New York: John Wiley and Sons, 1964.

Arnoff, F., Rubinstein, E., & Speisman, J. (Eds.), *Manpower for Mental Health.* Chicago: Aldine Publishing Co., 1969.

Broskowski, A., "Clinical psychology: A research and development model." *Professional Psychology*, 1971, 2, 235-242.

Broskowski, A. "Evaluation of consultation training programs." Paper presented at the American Psychological Association, 81st Annual Convention, Montreal, Canada, 1973.

Broskowski, A., & Baker, F. *Organizational and Social Barriers to Primary Prevention: An Ounce of Prevention May Cost a Pound of Cure.* Unpublished manuscript, Laboratory of Community Psychiatry, Boston, 1972.

Broskowski, A., & Khajavi, F. "Alumni of the Harvard Laboratory of Community Psychiatry." *American Journal of Communtiy Psychology*, 1973, 1, 62-75.

Caplan, G. *Support Systems and Community Mental Health: Lectures on Concept Development.* New York: Behavioral Science Publications, 1973.

David, H. "A perspective on manpower theory and conceptualization." In Arnoff, F., Rubinstein, E., & Speisman, J. (Eds.), *Manpower for Mental Health*, Chicago: Aldine Publishing Company, 1969.

Davis, W. *A Plan for Higher Education in Human Services.* Portland, Maine: Human Services Development Institute, University of Maine, 1973.

Demone, H. W., & Harshbarger, D. "The planning and administration of human services." In H. C. Schulberg, F. Baker, and S. Roen, (Eds.), *Development in Human Services, Vol. I.* New York: Behavioral Publications, 1973.

Department of Health, Education, and Welfare, Social and Rehabilitation Service, *Integration of Human Services in HEW: An Evaluation of Services Integration Projects.* Washington, D.C.: U.S. Government Printing Office, 1972.

Department of Health, Education, and Welfare. *The Second Newman Report: National Policy and Higher Education.* Boston: Massachusetts Institute of Technology Press, 1974.

Department of Labor, *Dictionary of Occupational Titles, 3rd Ed.* Washington, D. C.: U.S. Government Printing Office, 1965.

Department of Labor, Bureau of Labor Statistics, *College Educated Workers, 1968-80, Bulletin 1676.* Washington, D. C.: U.S. Government Printing Office, 1970.

Department of Labor, *Manpower Report of the President: A Report on Manpower Requirements, Resources, Utilization, and Training.* Washington, D.C.: U.S. Government Printing Office, 1973.

Edwards, C. "A candid look at health manpower problems." Speech to the 84th Annual Meeting of the Association of American Medical Colleges, Washington, D. C., November, 1973.

Fine, S. A., & Wiley, W. W. *An Introduction to Functional Job Analysis.* Kalamazoo, Michigan: The W. E. Upjohn Institute for Employment Research, 1971.

Galbraith, K. *Economics and the Public Purpose.* Boston: Houghton Mifflin Co., 1973.

Gross, B. M. *The Managing of Organizations.* New York: The Free Press, 1964.

Gruber, A. "The high cost of delivering services." *Social Work*, 1973, 18, 33-40.

Holden, C. "Higher education: Study urges altered thrust in federal support." *Science*, 1973, 182, 697-698.

Ingalls, J. *A Trainer's Guide to Andragogy. Revised Edition.* Washington, D. C.: U.S. Government Printing Office, 1972.

Klarman, H. "Economic aspects of mental health manpower." In Arnoff, F., Rubinstein, E., & Speisman, J. (Eds.), *Manpower for Mental Health.* Chicago: Aldine Publishing Co., 1969.

Knowles, M. *Modern Practice of Adult Education.* New York: Association Press, 1970.

Lanyon, R. I. "Mental Health Technology." *American Psychologist*, 1971, 26, 1071-1076.

Lewis, R. E., Anderson, J. A., Christiansen, J. H., Cooper, D. E., Gilbert, J. E., Henderson, K. M., & McShane, D. C. *A Systems Approach to Manpower Utilization and Training*. Salt Lake City: Utah Division of Family Services, 1972.

Likert, R. *The Human Organization: Its Management and Values*. New York: McGraw-Hill, 1967.

Marsh, T. W. *Manpower for the Human Services; Monograph No. 6, A Human Services Generalist Classification Series*. Chicago: Human Services Manpower Career Center, 1973.

McPheeters, H. L. *Roles and Functions for Mental Health Workers*. Atlanta, Georgia: Southern Regional Education Board, 1969.

Nadler, L. *Developing Human Resources*, Houston: Gulf Publishing Co., 1970.

Nathan, R. "Social policy in the 1970's." Paper presented at the Alumni Conference of the Florence Heller Graduate School for Advanced Studies in Social Welfare, 1973, Mimeo available from R. Nathan, The Brookings Institution, Washington, D. C.

National Institute of Mental Health, *Project Summaries of Experiments in Mental Health Training*. Washington, D. C.: U.S. Government Printing Office, 1971.

Putnam, J. & Chismore, W. *Standard Terminology for Curriculum and Instruction in Local and State School Systems; Handbook VI*. Washington, D. C.: National Center for Educational Statistics, U. S. Government Printing Office, 1970.

Rosenthal, G. "The economics of human services." In H. C. Schulberg & F. Baker (Eds.). *Developments in Human Services, Vol. II*. New York: Behavioral Publications, 1974.

Rostker, B. *An Econometric Model for the Evaluation of Manpower Programs*. Santa Monica, California: The Rand Corporation, 1973.

Schultz, T. W. *The Economic Value of Education*. New York: Columbia University Press, 1963.

Suchman, E. *Evaluative Research: Principles and Practices in Public Service and Social Action Programs*. New York: Russell Sage Foundation, 1967.

Tracey, W. R. *Evaluating Training and Development Systems*. New York: American Management Association, 1968.

Vidaver, R. M. "Developments in human services education and manpower." In H. C. Schulberg, F. Baker, and S. Roen, (Eds.), *Developments in Human Services, Vol. I*. New York: Behavioral Publications, 1973.

Weiss, C. H. *Evaluation Research: Methods of Assessing Program Effectiveness*. Englewood Cliffs, New Jersey: Prentice-Hall, Inc., 1972.

Williams, R. H. *Perspectives in the Field of Mental Health*. Washington, D. C.: National Institute of Mental Health, U.S. Government Printing Office, 1972.

Young, C. E., & True, J. E. "The current status of the associates degree movement in mental health and the human services." Paper presented at the American Psychological Association, 81st Annual Convention, Montreal, Canada, 1973.

Zytowski, D. E. *Vocational Behavior: Readings in Theory and Research*. New York: Holt, Rinehart, and Winston, 1968.

The key theme of this book, the integration of behavior analysis and systems analysis, is exemplified in the following chapter by Malott.[1] *Beginning with a brief description of those behavioral principles which offer promise in dealing with the management of complex social systems, and following Ford's suggestion*[2] *that we speak of "behavioral health" instead of "mental health", this chapter offers a conceptual background and practical suggestions for the planning, execution and maintenance of effective delivery systems.*

If well-executed, the behaviorally-oriented, systems-design approach facilitates healthy and productive activity on the part of both employees and clients. Several stages can be delineated in the design of an effective system. According to Malott, optimal results are obtained when a system-design is composed of: a careful preliminary analysis; concrete goal specifications; a behavioral design which includes specification, observation and consequation; reasonable implementation strategies; continual evaluation; and procedures which lead to a constant recycling through the entire process. In addition, the ideas contained in the concept of "total performance systems" help intervenors evaluate the effects of behavior change programs and, frequently, lead to viable management strategies.

In general, training and intervention will be effective only to the extent that the system is designed to insure that reinforcement is contingent on desired behaviors. Even for staff, appropriate behavior may not be its own reward and, if left to chance, contingencies often produce an unwanted product. Malott also suggests that managers and trainers be aware that initially appropriate behavior frequently deteriorates over time unless the system is designed to maintain it.

Richard W. Malott

CHAPTER 15

A Behavioral-Systems Approach to the Design of Human Services

The application of behavioral and systems analysis to the field of human services offers great promise. The approach outlined in this paper is more appropriately titled, behavioral health delivery systems. This perspective differs in design and procedure but encompasses such familiar areas as teaching, marriage counseling, prison reform, social work, etc., and offers insights into dealing with these problems.

BEHAVIOR ANALYSIS

In recent years, considerable evidence has been accumulated to indicate that it is both theoretically sound, and practically expedient, to consider human psychological problems as behavioral problems. In the past, psychologists have attempted to deal with inferred mental causes of behavioral problems, such as, "neurosis", "psychosis", "anxiety", "insecurities", etc. Though there is still room for improvement, we have been more successful since we have begun concentrating on the problem behaviors themselves rather than such inferred causes. Now we deal directly with overeating, poor work behavior, poor study behavior, excessive talking, etc. Our success with such behaviors prompts us to persist in a behavioral approach to less tractable problems such as excessive drinking, drug-taking, and marital conflict.

Behavioral Health

A position advocated by Ford (1974), and in keeping with the behavioristic philosophy of this work, suggests that we stop talking about "*mental health*" and speak instead of "*behavioral health*". While we might retain the term "mental" for historical reasons and argue that its old-fashioned implications are essentially benign, our orientation might be more accurately described by the term "behavioral health". The latter term more correctly reflects our analyses and procedures.

Ford has also suggested that we define "behavioral health" in terms of the goals of our general society and not in terms of the happiness of the individual members of that society. For example, in the education of scientists, our objectives are not to produce happy scientists but rather to produce productive scientists. To the presumably great extent that happiness, however defined, facilitates the productivity of the scientists, then happiness can be considered a part of our objectives.

One of the great historical and philosophical debates has raged over the question of the relative priorities of the rights of the individual as opposed to the rights of the larger system, society, or culture. However, it is most generally the case that behavioral-health delivery systems are concerned with the "happiness" of the individual, as well as his productivity. Social environments can be constructed in which individual productivity and happiness are not sacrificed to the larger system, as is often the case under the haphazardly designed political, social, and economic systems in today's societies.

Behavioral Consequences

One of the most outstanding features of behavior is that it is controlled primarily by its consequences. Some behavioral consequences increase the likelihood of future occurrences of that behavior, others decrease that likelihood.

Consequences that increase the likelihood that a behavior will occur again are called *reinforcers.* These reinforcers are determined by existing conditions and change as conditions change. For example, a cold bottle of beer might be a very effective *reinforcer* for the behavior of a person who has been without something to drink for 24 hours. By the same token, for the individual who has consumed several bottles of beer, the opportunity to go to the bathroom will be a very powerful reinforcer.

Consequences that *decrease* the likelihood of future occurrence of the behavior they follow are called *punishers.* If beer drinking leaves the individual with a hangover, then the consequences of drinking the beer will be punishing. Similarly, if there is no bathroom available to someone who has drunk several bottles of beer, those consequences will be punishing.

The basic task of the behavioral-health worker is to help arrange environments so that reinforcing consequences follow appropriate behavior, and no consequences, or punishing consequences, follow inappropriate behavior — a task easy to state but often difficult to accomplish.

Conditioned Consequences

Because many of the consequences of human behavior are extremely subtle, they are difficult to detect and analyze. One of the main difficulties in dealing with these consequences lies then in determining just what they are, and whether they are, in fact, reinforcing or punishing.

Aside from unlearned consequences such as a reinforcing beer, or a punishing hangover, there are many consequences that are learned or *conditioned* — they acquire their value through association with other consequences. For example, the very powerful conditioned reinforcer of attention may acquire its reinforcing value because attention has been intimately associated with past receipt of the wide range of reinforcers which come from other people.

The difficulty in determining the value of a consequence becomes apparent in the all-to-common situation in which a reprimand for inappropriate behavior, may, in actuality, serve as the reinforcing consequence for that behavior. The reprimand may constitute the only significant attention the person in authority offers the individual exhibiting deviant behavior. Ironically, attention in such cases works as a detriment to the goal. Such reprimands are often demonstrated to be the major source of reinforcing consequences, thus serving to maintain deviant behavior rather than reduce it.

Immediate vs. Delayed Consequences

Another feature of behavior that must be dealt with is the fact that behavior is much more strongly influenced by immediate and certain consequences than by delayed and less certain ones. For example, suppose your husband or wife mildly offends you. Under the circumstances, it will be slightly reinforcing to say something nasty to your spouse. And trivial though that reinforcer for nastiness may be, it is very immediate and very definite.

It would be better to ignore the offense or to deal with it in a much less sarcastic manner. Unfortunately, the consequences for this more desirable response are much too long range, uncertain and diverse. Even though these distant consequences may be more important — a prolonged and happier marriage — they exert much less control on your behavior than the immediate reinforcers for a sarcastic retort. Many of our behavioral problems are due to this very real fact that trivial, but immediate, behavioral consequences have a greater influence on our behavior than do more important, but distant and uncertain, effects.

Immediate vs. Historical Environment

We must also recognize the fact that the immediate environment generally exerts much greater control over behavior than do past environments. If we wish to affect human behavior, we must deal with the ongoing environment and not rely on historical variables to generate proper behavior.

For example, many therapeutic procedures are based primarily on the notion that the experiences the patient undergoes in a therapy situation will cause him to behave in a healthy manner at a later date, and in a different place. Such procedures will not be nearly as successful as procedures that are designed to directly modify the future environment of the patient. Similarly, attempts to train children or adults to be good students will not be nearly as effective as procedures which set up appropriate reinforcement contingencies for study behavior in the future environment.

Change the environment — not the person. Change the classroom — not the student. If you wish to increase the amount of appropriate study behavior in the classroom, arrange for the teacher to set up reinforcement procedures for that behavior. Don't spend an undo amount of time trying to convince the students that they really want to study, that if they do study they will be happier adults.

Resistance to Extinction

In a similar vein, some behaviorists first establish appropriate behavior through the reinforcement of every occurrence of the desired response, and then gradually reduce the frequency of that reinforcement. They reason that by using occasional or intermittent reinforcement, they will be able to greatly increase that desired behavior's resistance to extinction once the individual has been removed from the program, and is back in the normal environment.

One of the factors that makes this unworkable is that in heterogeneous human social environments there will be competing sources of concurrent reinforcement for inappropriate behavior. When reinforcement for the appropriate behavior is completely removed its rate of occurrence will probably decelerate much more rapidly just because of this competition.

Institutional and Natural Environment

There are two general environments in which the behavioral-health professional will perform — institutional and non-institutional (i.e., "natural") settings. Modifying behavior is not an easy task in either setting. However, the institutional environment lends itself more readily to this undertaking.

Hospitals, schools, sheltered workshops, prisons, day-care centers, etc., provide the most ideal background because they offer two particularly useful features that facilitate the design of a behavioral procedure and enable the behavioral-health professional to best accomplish his goal. First, they are already staffed with personnel whose task it is to deal with the behavior of the inhabitants. This is an invaluable asset as in setting up behavioral procedures, it is usually necessary to have an individual who observes and consequates behavior. The institutional staff frequently serves this function.

The second important feature of institutional settings is that they usually occupy a relatively small amount of space. It is much easier to observe the behavior of the child in the classroom than it is to monitor that behavior in the wide variety of locations he will inhabit outside of school. It is not surprising, therefore, that most of the advances, to date, in behavioral technology have been restricted to institutional settings. Only recently has behavioral work been instituted in the area of non-institutional or natural environments.

The managerial difficulty of modifying behavior in the natural environment is primarily the matter of control. Consider the following example: An adult client wants to quit smoking. Behaviorally, this is a relatively easy problem to deal with: We observe the individual each time he takes a puff of a cigarette and apply some mildly punishing consequences immediately following each response. If the individual is confined to a limited environment where his behavior may be constantly observed, this procedure can be readily implemented. However, constantly observing an individual's behavior in the wide variety of settings normally available to the adult is an extremely difficult managerial problem.

A SYSTEMS-DESIGN APPROACH

The task of the behavioral-health worker is to help establish a behavioral system to facilitate healthy behavior on the part of the individuals in that system. Frequently, the system may be quite small, for example, the members of a family; nonetheless, a systems-design approach can be followed. It is my contention that the design of any system be it mechanical, electrical, biological, or behavioral will proceed best if a set of general, systems-design guidelines are followed.

The basic systems-design concepts are very powerful and are generally independent of the nature of the specific system being considered. The six basic stages of systems design are as follows: *preliminary analysis* of existing systems, or of the problem area; *specification of the goals* of the system; *design* of the system; *implementation* of the new system; *evaluation* of the new system; *recycling* through the previous stages of systems design until a system has evolved that accomplishes the specific goals.

While these stages may seem obvious, most behavioral or social systems are not the result of such a systematic design process and should profit greatly from such an approach. Attempts to cut short this process when dealing with a single individual or a small family will generally result in an inadequate behavioral system that will not survive. It is essential to go through all six phases: preliminary analysis, goal specification, design, implementation, evaluation, and recycling.

Using a combination of a behavioral and a systems-design approach let us briefly consider how a behavioral-health worker might function as a marriage counselor in dealing with the aforementioned behavior commonly involved in marriage difficulties.

Preliminary Analysis

One of the most frequent problems in a marriage is the high rate of negative, aversive verbal behavior between the partners. As problems develop within the union, it becomes reinforcing for each partner to make negative, sarcastic and critical remarks to and about the other partner. Immediate reinforcers gain excessive control of their verbal behavior to the detriment of long-range objectives, those of happiness and compatibility.

Goal Specification

A behavioral objective in designing an improved marriage system, then, might be to reduce negative verbal behavior to a tolerable limit. For example, each partner might agree to make no more than two negative remarks per day. With this goal in mind we can move into the design phase.

Design

In designing a behavioral system, it is useful to divide the phase into three steps: *specification*, *observation*, and *consequation*. You should always precisely *specify* the behavior under consideration. Next you should arrange to have the behavior *observed*. Finally, arrangements should be made to have reinforcing or punishing *consequences* occur immediately following that behavior. Failure to deal with each of these three aspects of a well-designed behavioral system will generally result in a weak system.

In an effort to help the couple become more aware of their negative verbal behavior, they might be instructed to record (*observation*) each instance of their own and of their spouse's negative verbal behavior (*specification)*, then to plot those results on a graph at the end of each day (*consequation*). In many cases simply recording one's own inappropriate behavior results in that behavior decreasing to an acceptable level.

Implementation

A common problem with good advice, even from professional marriage counselors, is that people may only rarely follow it. It is essential to be certain that the newly-designed system is actually implemented.

Evaluation

Specific recommendations of mental or behavioral-health workers are essential in the development of an improved system. Unfortunately, it is quite rare that their recommendations are systematically evaluated as to their effects on behavior. It is, of course, much easier to give advice than to evaluate the efficacy of that advice. However, only through close evaluation can we determine if our program is working. Further, if our design isn't providing the desired results, careful evaluation can reveal the weak spots in our overall plan.

Evaluation may show, for example, that the married couple are still complaining — still exhibiting a high rate of negative verbal behavior, and perhaps doing an inadequate job of recording their behavior.

Recycling

Particularly when dealing with a new problem, the behavioral-health worker should anticipate recycling through the systems-design process several times before the desired results are achieved. Although the behavioral-health worker may have to institute a variety of changes and recycle several times, this procedure, if properly and consistently instituted, will ultimately insure the desired end result.

In recycling, several changes might be deemed appropriate. Initially, a goal of only two negative responses per day may prove to be unrealistically low. Perhaps external or extrinsic consequences might be used to provide specific reinforcers when the rate of negativity is low, and a specific punisher when the rate of negativity is high. It may be necessary to more definitively specify what constitutes negative behavior by listing a large number of examples of such behavior.

Perhaps some sort of consequence should be provided for recording, and failing to record negative behavior throughout the day. In this case it may be necessary to design an adequate system to accomplish this. In the interest of simplicity the couple might consider a recording system such as a golf counter. This can be worn on the wrist to increase the facility with which each records his or her negative behavior.

The skeptical reader would need many more examples to be convinced; nevertheless, I assert that the six general phases of systems-design, combined with the three behavioral steps in the design phase, are quite general, and applicable to essentially any behavioral-health problem and solution.

We have looked at some important features of human behavior as well as a preliminary analysis of some of the factors which lead to poor behavioral health. In addition, we have assigned to the treatment system the goal of improving behavioral health. As we have stated, the general goal of such a system is to generate a high level of behavioral health in the community. We have also loosely defined healthy behavior as productive behavior, acknowledging that there is still considerable room for a greater degree of specificity.

This constitutes our general model for dealing with behavioral-health problems and their treatment. The next question is: How do we design behavioral-health delivery systems that will maintain good behavioral health and be capable of treating behavioral problems when they appear? What will a well-designed behavioral-health delivery system look like?

STAFF PROGRAMS

In designing systems to train and maintain appropriate behavior on the part of institutional staff, we should use the same general systems-design procedures and behavioral-analysis principles specified earlier.

We should never assume that appropriate behavior is its own reward; learning is not its own reward, teaching is not its own reward, increasing behavioral health is not its own reward. A systems-design is incomplete until explicit reinforcers and punishers are made contingent upon the appropriate and inappropriate behavior of every individual in that system. Failure to do this will definitely jeopardize the survival of the system by allowing the haphazard contingencies of the undesigned system to maintain much inappropriate behavior.

Staff Training

While a staff training program will not be sufficient to guarantee the *maintenance* of a behaviorally-designed program once it has been implemented, such a training program will greatly facilitate its implementation. An understanding of applied behavioral analysis and behavioral-systems design on the part of the staff will not only provide the basis for their use of behavioral procedures with the occupants of the institution, but will also smooth the way for their acceptance of a behavioral system designed to maintain their own mediating behavior. It is my feeling that the miracle of a successful behavioral system is usually not the fact that it works but that the designers were able to implement and maintain it. There is many a slip between the design and implementation.

Reinforcement

In order to have a consistently effective training program, it will be necessary to use some sort of extrinsic reinforcement for the staff participation in this program. For example, if outside reading is expected, then it will be necessary to observe and consequate that behavior. Do not make the error of assuming that because the staff *should* participate actively in the training, and because they are mature and responsible people, that they *will*, in fact, do so.

If participation in the training program, itself, is a reinforcer, then it may be possible to make continued participation in the program contingent upon mastery of behavioral objectives along the way.

The staff may find college credit (graduate or undergraduate) a powerful reinforcer. Under those circumstances, it may be possible to arrange for the staff to enroll in the training program for course credit. This procedure has been used successfully many times. Incidently, an in-service staff training program of this type is probably the ideal educational setting, as it allows the student to integrate theory and practice to a degree which only rarely occurs in the typical college classroom. Money is also an ideal reinforcer for maintaining active participation in a staff training program; however, this reinforcer rarely seems to be available for such purposes.

Ubiquitous Training Topics

The successful implementation of a behavioral system requires skill at public relations as well as a level of interpersonal sensitivity not normally associated with behavioral psychologists. In addition to encountering the cultural inertia met by any new program, the behaviorists must deal gently but effectively with many common-sense, psuedo-moralistic issues before the system can be successfully implemented.

One of the most ubiquitous of stumbling blocks is the fact that the use of extrinsic reinforcers, for normal behavior, runs counter to our intuitive morality. If a person is not studying properly, not eating properly, not working properly, not interacting with others properly, there is a strong cultural bias to punish the maladaptive behavior rather than to reinforce appropriate or normal behavior. "Sally does her homework without any special rewards, so it's not right to give Johnny special reinforcers, 'bribes', to get him to do what he should do anyhow." Successful implementation will be greatly facilitated by first dealing adequately with these common objections as you move into your program.

Staff Maintenance

In recent years we have learned a great deal about which behavioral programs work best in maintaining the appropriate behavior of individuals in institutional settings. Now the major problem becomes the determination or design of optimal strategies for implementing those behavioral programs. For example, we can easily design behavioral systems that will maintain a high amount of productive study behavior on the part of children in the grade school classroom. The major problem is to get those designs implemented in the school system; and once implemented, to have them maintained.

Most behavioral procedures require a certain amount of immediate effort on the part of the staff responsible for the procedures. Without designing a system to maintain the behavior of the institutional staff, the behavioral systems designed for the occupants of the institutions generally deteriorate.

Even after successful behavioral teaching procedures have been implemented in the elementary school classroom, some system of extrinsic reinforcement must be established to maintain the behavior relative to those agreed-upon procedures. The teacher's behavior is quite likely to fall under the control of immediate consequences that are incompatible with the originally agreed-upon procedures. This, of course, is equally true of staff in other institutions: attendents in mental hospitals, staff in sheltered workshops, and college instructors. . .

In dealing with behavioral health, we have indicated that more emphasis must be placed on the impact of the immediate environment upon immediate behavior, and less reliance should be placed on the use of the immediate environment to influence future behavior. Similarly, training programs attempting to influence the future behavior of institutional staff are of limited value. They must be part of a total design entailing continued reinforcement for appropriate staff behavior throughout the future existence of that institution.

Staff as Mediators

In institutional settings, the staff who deal directly with the observation and consequation of behavior may be called "mediators". They mediate between the designers of the behavioral system and the behavior under concern. For that reason it is essential also to observe and consequate that mediating behavior. This applies equally to the natural environment.

In parent counseling, the parent may be trained to be the "mediator" in dealing with the child's behavior. Under these circumstances, one should not assume that a well-behaved child will be a sufficient reinforcer to maintain appropriate observing and consequating behavior (i.e., mediating behavior, on the part of the parents). It is as necessary to design a system that will provide observation and consequation for the parents' mediating behavior as it is to design one for the child's behavior.

Staff Meetings

Once the staff-training program is completed and the system is implemented it may be appropriate for the trainers to shift into the role of evaluator, monitor, maintainer of the staff's mediating behavior. One way to accomplish this monitoring is to have the staff continuously collect data assessing the effectiveness of the procedures they are using. For example, a grade school teacher might collect data on the study behavior of all of the class or of certain problem children. Or he might choose to measure behavioral progress through the use of frequent reading tests.

Each staff member would then present his data at periodic staff meetings. The social reinforcement involved in peer and supervisor approval may be sufficient to maintain the staff's attending behavior. It may also maintain presentation of adequate data reflecting effective performance of the regular daily duties in the behavioral program. Although periodic staff meetings are frequently rather dull, they are very important to the maintenance of a healthy and positive community.

If the norms of the staff community are such that attendance becomes sporadic, then it may be necessary to arrange extrinsic reinforcing consequences for attending the meetings and presenting behavioral data and records. (One possible system might be to provide a small amount of academic credit for regular participation, so that a student/staff member who attended most of the weekly meetings for a year would receive academic credit equivalent to a typical one-semester class.) In any event, if extrinsic reinforcers, successful in getting the staff through the training program, were found, those same reinforcers would probably function well in maintaining the staff's subsequent mediating behavior.

THE TOTAL PERFORMANCE SYSTEM

Brethower (1972) presents a set of concepts that are helpful in the analysis of systems. He calls this a *total performance system* — a human performance system which, when operational, functions to improve and/or maintain its performance. He divides this total performance system into two major components: a *processing system*, in which one or more inputs are changed into one or more outputs; and a *receiving system*, which receives the outputs of the processing system.

A behavioral-health program such as a hospital or clinic could be considered a processing system, while the patients, their friends, family and employers constitute part of the receiving system. Patients themselves serve as part of both input and output — entering patients are part of the *input*, and patients completing the program, part of the *output*.

Looking at behavioral-health programs as a total performance system emphasizes the fact that the behavioral-health program designer must be concerned not only with the processing system (i.e., the hospital or clinic), but also with the receiving system, the patient's world once he leaves the institution. Sadly, there has been very little systematic analysis of the receiving system.

Within this context, both objectives and goals constitute the desirable outcomes of the system. Objectives are measurable, short-range performance outcomes which the system-designer specifies and implements. Goals are the idealistic but workable, eventual outcomes of the processing system.

For instance, typical objectives for a prison system might be the inmates' improved cleanliness, more productive work activities, better performance in an educational program, and development of appropriate verbal behavior. The goals of such a system might be the return of the inmates to society where they would successfully deal with that environment.

These total performance systems concepts will prove most useful in analyzing the behavior of the systems-designer and/or manager.

THE SYSTEMS-DESIGNER AND MANAGER

And what about the systems-designer? Who observes and consequates his behavior? Under ideal circumstances, the systems-designer will obtain his ultimate reinforcers from the receiving system for designing the processing system. Even more idealistically, those reinforcers will be dependent upon the extent to which his system-design accomplishes its objectives.

Accountability

Attempts have been made to include this sort of accountability in the design of recent elementary-educational systems called "contract education". In this innovative design the contractor and the receiving system (school board) agree upon a clearly specified set of educational objectives; for example, the attainment of certain reading levels by students in a particular class. The contractor's salary is then dependent upon the extent to which he accomplishes the contracted behavioral objectives.

This sort of accountability will see increasing use in other areas of human services. The systems-designer who is serious about accomplishing the goals of the system is well-advised to include, within his design, a procedure for the observation and consequation of his own behavior.

Accountability to the Profession

Their profession serves as a major source of natural consequation for many systems-designers. Reinforcement for professional behavior is provided by the opportunity to publish articles in professional journals and present papers at professional meetings. There is a high probability that the main function of these meetings is not to act as a source of new information for the participants but rather to provide the participants with an opportunity to receive social reinforcement for the professional activities upon which they report — a professional show and tell.

However, the professional group, as a receiving system, does not always have the same goals as the receiving system which pays the salary of the systems-designer. Consequently, the systems-designer may devote a considerable amount of the system's resources to the performance of "scientific" experiments or other activities that will be reinforced by a professional community but not by his employers.

Manager Maintenance

Where you find an effective behavioral system in operation, you will find a behavioral systems-designer and manager primarily responsible for keeping the system operative and on target: the direction of his behavior is greatly under the control of reinforcers from a professional audience. Although this audience reinforces innovative design behavior, it provides little reinforcement for the maintenance of the novel behavioral system once it has been established. In addition, unless it is specifically designed into the system, gratitude and other forms of social and material reinforcers for the manager from a grateful receiving system are rarely forthcoming.

A system for the continuous maintenance of a capable manager must be included in any successful design. When the original designer stops receiving enough reinforcement to sustain his participation in that system, he moves on to other projects. The consequence is that many well-conceived systems have gradually deteriorated without the continued guidance of the designer.

EVALUATION OF RECEIVING AND PROCESSING SYSTEM FEEDBACK

Receiving System

It is frequently true that the first design of a complex behavioral system is inadequate to the task of accomplishing the objectives of the system. It is almost always necessary to recycle through the design process on the basis of an evaluation. Therefore, a thorough evaluation system is essential.

A well-designed system should have a set of objectives and goals agreed upon by the designer, the individuals in the processing system, and the individuals in the receiving system. A common flaw, however, in designing behavioral systems is that rarely are there adequate resources allocated for the evaluation phase. Without this systematic evaluation and recycling, few systems will accomplish their objectives.

Ideally there should be two independent sources of evaluation — the receiving system and the processing system. The main purpose of an evaluation by the receiving system is to determine whether or not to continue to financially support the processing system. The receiving system should evaluate the effectiveness of the processing system, both by looking at the feedback from the processing system and by examining feedback from the receiving system itself.

However, very often evaluation by the receiving systems tends to be concerned primarily with receiving-systems' feedback and less with the operations of the processing system. It might be a good idea in the original design of the total performance system to provide for the receiving system to hire a professional evaluator — someone who is trained in the evaluation of behavioral-health delivery systems. He might report to a panel composed of members of the receiving system who would then make the final decision as to whether or not to renew funding for the processing system.

It's important to design procedures that prevent the professional evaluator from being too strongly influenced by his professional colleagues in the processing system, otherwise the "neutral" evaluator may find himself involved in a conflict of interest.

Processing System

On the other hand, the designer and the processing system should evaluate the feedback from both the processing system and the receiving system in order to effectively recycle through the design process to improve the behavioral system. It is important that the evaluation of this feedback be quite candid. Therefore, it is appropriate that the processing system evaluation be kept confidential. If this evaluation is also to be used by the receiving system to determine whether or not to continue to support the program, there will be a strong conflict of interest. This will result in the evaluation being much more positive than it should be in order to insure further support of the program.

If the designers are unable to arrange for an evaluation by the receiving systems, they should arrange for at least part of the evaluation prepared by the processing system to be made available to the receiving system. This should be done carefully so as not to bias the evaluation which the designer uses in recycling through the design procedure. This is vital, for without systematic evaluation and recycling few systems will ever accomplish their objectives.

Convenience Factor

Even though there may be an explicitly stated set of objectives, there is probably also an implicit set of objectives equivalent to the "hidden agenda". The individuals in the processing system may behave as if their primary concern were their own convenience and expedience, even though these are not explicitly stated as objectives. Without continued monitoring by the receiving systems, the processing system may gradually drift into a program that is designed to make the lives of the staff easier, but not necessarily designed to accomplish the originally agreed-upon behavioral-health goals.

For example, in classrooms, mental hospitals, and other institutional settings, there is a strong tendency on the part of the staff to use behavioral, pharmacological and even surgical procedures that are designed to minimize disruptive behavior of their wards, thereby facilitating the tasks of the staff. These are worthy objectives and may also be necessary in order to obtain other long-range, behavioral-health objectives beneficial to the students or residents.

However, in other cases the results are highly questionable. With procedures such as frontal lobotomies, the attainment of an easily-managed resident may create an individual who will be much less likely to acquire healthy, appropriate behavior patterns necessary for non-institutional living.

Even the less extreme practices utilizing behavioral procedures are very susceptible to this. A classroom or hospital environment in which those in charge reinforce silent docile behaviors will not generate the kind of creative and productive behaviors necessary for successfully dealing with the natural environment.

If given a choice between a procedure that will be most convenient for the staff, but may have a negative therapeutic value, and a procedure that is quite inconvenient for the staff but will probably have a powerful therapeutic value, the staff may have a strong tendency to select the former and, with little difficulty, develop socially acceptable rationalizations. This is true of any staff be they paraprofessional hospital attendants or professional nurses and psychologists.

Recycling

It may also be desirable to have an ombudsman within the processing system. The ombudsman would receive processing system feedback from individual patients, students, residents, or staff members, and then channel that feedback to the proper recipient within the processing system. A neutral, honest, and respected mediator of this type can be a powerful influence in re-ordering the processing system. Negative feedback from patients, residents or students is much more likely to be effectively received when channeled through such an individual.

Evaluation, of course, is not an end in itself, simply the final step taken in the systems-design cycle before the designer recycles to the beginning in an effort to redesign his systems to more readily accomplish the agreed-upon goals and objectives.

His first step in recycling through the design process should be to reanalyze the behavioral processes involved: were there important controlling variables in effect that had been overlooked during the preliminary analysis?

During the next phase, the designer should reexamine the previously specified objectives and goals. On the basis of the first cycle through the design system, does it appear that it will be feasible to accomplish all of the objectives? If not, this would be a good time to reassess the list of objectives in order to bring them a little closer to reality. Some of the objectives may prove to have been vague and ambiguous and need to be stated more explicitly.

Finally, the system should be redesigned, implemented, evaluated, and, if necessary, continuously recycled until the system performance matches the specified objectives. It is probably the case that in a real-world system, the recycling process will be a continuous operation; we never get there, we just get a little closer.

COST-EFFECTIVE BEHAVIOR-HEALTH SYSTEMS

In designing any system, one should attempt to maximize the benefits derived from the system while minimizing the costs. This sort of cost-effectiveness analysis has only recently been applied to human services programs.

Goals and Objectives vs. Activities and Structures

Such cost analyses are facilitated by defining programs in terms of specific goals and objectives rather than activities and structure. If you define a community mental health program as consisting of an office, two clinical psychologists, two social workers, and a secretary, then the notion of cost-effectiveness is not too applicable. You pay the rent and salaries, and that is it.

On the other hand, if you define a community mental health program in terms of goals and objectives, then the task of the designer is much more complex than merely hiring the standard staff and renting the standard space. After he selects a set of objectives for the program, the designer should consider alternative ways of accomplishing those objectives. It is in this way that he or she will probably have the most significant impact on increasing the cost-effectiveness of the program.

Goal orientation is desirable as it will greatly free the designer to develop creative solutions to the achievements of those goals. A task, activity, or structural orientation, on the other hand, will impose undue constraints on accomplishing anything except the replication of those traditionally determined tasks, activities, and structures.

Marriage Counseling

If a marriage counselor defines his services as increasing the healthy marital behavior of his clients, then a great number of activities might be available to aid him in accomplishing those objectives. He might arrange for the couple to develop behavioral contracts; work in their home setting with outside mediators; record their own behavior; take part in feedback sessions based on their behavior in the counseling clinic; as well as indulging in the more traditional advice-giving role of the marriage counselor.

The traditional counselor will restrict himself to the more traditional activities of trying to eliminate marital problems once they have threatened the marriage; however, the goal-oriented marriage counselor might also design programs for happily married couples in order to prevent the development of such marital disharmony.

Job Placement

It is virtually impossible to develop new and more effective programs if the programs are constrained to traditional tasks. However, creativity virtually flourishes when the analysis is based on goal-oriented objectives. An excellent example of this is the effort by Jones and Azrin (1973) to develop a job-placement program for past patients of a mental hospital. They did not define their services in terms of traditional professional activities but rather in terms of accomplishing the objectives of gaining employment for the former patients.

Their preliminary analysis of the *real* job placement system indicated that jobs were more often obtained through the candidate's acquaintances than through formal job placement services. In other words, the people who are actually responsible for the job placements were receiving extrinsic social reinforcement for those activities.

Suspecting that the patients of a hospital are not in a position to provide such social reinforcement, Jones and Azrin decided to add some explicit, extrinsic reinforcement of their own. They advertised the fact that a one-hundred-dollar reward would be given to individuals responsible for placing people from their program. Consequently, their alumni experienced markedly less difficulty in getting jobs.

Budgets

One major obstacle in the development of innovative programs is that budgets are frequently tied to activities and structure rather than objectives. It is doubtful that there is an existing category for job-finder's fees. This problem of the inflexible budget is all too frequently encountered in attempts to establish meaningful monetary reinforcers to be made contingent upon improved staff performance. Developing project budgets in terms of goals and objectives would greatly facilitate cost-effective accomplishment of those objectives.

Maximizing Effectiveness

The concern for cost-effectiveness is not necessarily an outgrowth of a desire to reduce the overall cost of the program so that money can be returned to the taxpayers. Instead, it is generally the case that the program has a fairly limited budget, and it becomes the designer's job to maximize the benefits that can be derived from that budget. This may be done by reducing the cost of many of the individual services, or components, and re-allocating resources to increase the effectiveness of the general program by providing more services or better services.

However, this does not mean that we should ignore the possibility of total cost reduction where applicable.

The fixed-income nature of most government-supported programs discourages attempts to decrease the overall cost of the program. In fact, there are cases in which those who try to reduce excessive cost are punished for their efforts. The recent incident of the Pentagon official who was fired as a result of calling attention to cost overruns exemplifies the failure of fixed-income systems to provide reinforcement for their monitoring of cost-effecitveness.

This characteristic of fixed-income systems may be one of the reasons that cost-effectiveness analyses have been somewhat late in appearing in publicly supported programs. There is certainly considerable incentive for reducing program costs in a profit-making environment.

Paraprofessionals

One of the more efficacious ways of improving the cost-effectiveness of a human services system is to make increased use of paraprofessionals, people who have considerably less training than that normally required for professional positions. Examples of paraprofessionals are teacher's aids and ward attendants.

Programs that are not evaluated on the basis of the accomplishment of behavioral objectives may attempt to attain prestige by pointing to the high professional qualifications of their staff. Such criteria are not conducive to the effective use of paraprofessionals. For example, a community mental health center exhibiting this tendency might be much more inclined to hire an additional clinical psychologist than three paraprofessional case workers even though the additional, though not highly trained manpower provided by the three paraprofessionals might be much more efficient than the addition of a single clinical psychologist.

In any human services program, highly trained and expensive staff spend large portions of their time in activities that could be carried out virtually as well by much less expensive staff with considerably less training. For instance, many of the applied behavioral procedures which originally evolved as a part of master's and doctoral theses are now being effectively utilized by teachers, parents, and attendents who have had only a minimum of training. This can be particularly true if the paraprofessional staff is provided with in-service training that is specifically geared to the attainment of the program objectives.

Mass Production

Another approach to improving cost-effectiveness is the use of standardized, mass-produced devices to deal with problems that have a common solution. For example, in the classroom, textbooks and movies are used as instructional tools. It is much less expensive to have a student read a book for one hour, or to have twenty students look at a movie, than it is to have a single teacher work with a single student for that hour.

Similarly, in other behavioral-health programs, a large part of the preliminary treatment of various problems might be handled more efficiently if the patient or client simply watched a movie and/or read some material prior to extensive interaction with the staff. This eliminates the expensive and time-consuming preliminary oral orientation presented by a highly-paid staff member.

This is not meant to imply that the client, patient, or student will be consistently dealt with as a component module in a mass-production system. When more behavioral procedures are necessary to solve individual problems, more specific contingencies in the form of personal counseling or remedial help can be implemented. The value of mass production is that it improves the system by improving the behavioral health of more people at lower cost.

Computer Systems

A third area of increased cost-effectiveness in behavioral-health delivery systems will undoubtedly come from the use of computer terminals which interact with the individuals to help deliver the behavioral services. At present, the greatest amount of work in this area seems to be in the use of computers in education; however, they have yet to have any widespread impact.

Priorities

Once the most effective approach is selected for each of the objectives of the program, it will usually be found that there are still not enough resources to accomplish all of the specified objectives. Under these circumstances, the various objectives should be assigned values or priorities.

Minimize the Number of Objectives

Most human services programs dilute their efforts by attempting to simultaneously accomplish too many diverse objectives with inadequate resources. In many situations, doing the job only halfway is as unsatisfactory as failing to do it at all (i.e., a miss is as good as a mile).

For example, a weight-reduction program that helps people lose weight for five weeks, after which they rapidly regain their initial weight, is of little value. A program to deal with school truency that affects an individual child for only a few weeks is of little value. A program to train the retarded in self-help skills is of little value if it never gets beyond the prerequisite skills, and fails to establish behaviors that will be maintained once the preliminary training program has been terminated. While many of us become skilled at hypothesizing potential or theoritical values of incomplete programs, these hypotheses are probably only rationalizations for failure. Unless the program results in some measurable positive consequences, it should remain suspect.

Select Realistic Objectives

Another way to improve the effectiveness of the overall program is to place primary emphasis on those objectives that will have a reasonably high probability of being accomplished. A very common problem with human services is that the objectives may be selected solely on the basis of their importance, with little regard for the achievement probability. A program's overall effectiveness could be greatly increased by avoiding those projects that are virtually doomed to failure.

One common error is to implement programs with patients or clients in a natural environment where there will be no reliable day-to-day mediator available. For example, Tharp and Wetzel (1969) in designing programs for pre-delinquent juveniles found that it was essential to have a parent, teacher, or some other responsible mediator, in the child's environment in order to facilitate the observation and consequation of the child's behavior. If a reliable mediator could not be found, then the probability of success in dealing with the child's behavior was too low to justify the costs of implementing a behavioral program.

The cost-effectiveness of the program will be greatly increased if we select only the top few of the most important objectives to deal with; we should select only as many objectives as can realistically be accomplished with our pitifully limited budget. In so doing, it should be understood that all new projects cost twice as much and take twice as long as any reasonable person would have anticipated.

Consider Both Cost and Value

The actual selection of projects may be more complicated than simply selecting those with maximum priority, if one wishes to maximize cost-effectiveness. Perhaps the two highest-priority objectives would be exorbitantly expensive. Furthermore, maybe ten lower-priority objectives could be accomplished for the same cost. And suppose the total value of those ten lower-priority projects greatly exceeds the total value of the two maximum-priority projects. In other words, in selecting objectives for maximum cost-effectiveness, we must not only look at the value of accomplishing the objectives, but the cost of so doing.

While the selection of the optimum mix of objectives is undoubtedly amenable to mathematical programming, the actual assignments of value and costs to behavioral projects is sufficiently vague that the use of high-level, quantitative technology would be like gilding the dandelion. At this point in time the design of human service delivery systems is sufficiently primitive that the simple use of a preliminary cost-effective analysis on an intuitive basis should result in a major step forward.

The Follow-Up Maintenance Program

As we have indicated, in designing a set of procedures for the implementation of a behavioral program, it is desirable to begin with an in-service staff training program. It is also very important to provide for a systematic program to observe and consequate the staff's mediating behavior. In other words, it is not only necessary to train the staff, but it is also important to maintain the staff in their everyday activities once the training program is terminated.

A well-designed behavioral-health program will also have two phases: first, an intensive behavioral program to bring the appropriate behavior up to a desirable level and the inappropriate behavior down to an acceptable level; then a follow-up program to maintain those levels of behavior once they have been achieved. If a weight-reduction program has successfully helped a group of clients control their eating and exercising behavior so that they have attained the weights they desire, it is also necessary to implement a maintenance program to help the client maintain his weight.

Almost all human services programs fall way short of the mark concerning follow-up or maintenance procedures. Most programs would greatly enhance their effectiveness if they would cut the size of the regular program in half and put the resulting resources into maintenance programs.

SUMMARY

The recognition of the basic principles of human behavior, and the utilization of them in the design of behavioral-health delivery systems offers many advantages and possibilities. The basic, practical benefits of such systems are increased cost-effectiveness, better staff performance, and continued efficient maintenance of effective programs. The possibilities of the widespread use of these systems are not presently measurable. However, it is most evident that systems which are efficient and successful in producing students with better educations, improving the behavior of patients in mental hospitals, returning prisoners to society, and increasing the happiness and productivity of people in business and industry will be among man's most worthwhile efforts.

FOOTNOTES

[1] The author wishes to acknowledge the support of Western Michigan University in providing a sabbatical leave of absence during which this chapter was prepared.

[2] Presented at the West Virginia Conference — "Issues and New Developments in Mental Health Services", Morgantown, May 1973.

REFERENCES

Brethower, Dale M. *Behavior Analysis in Business and Industry: A Total Performance System.* Kalamazoo, Michigan: Behaviordelia, 1972.

Ford, Donald H. "Mental health and human development: An analysis of a dilemma." In D. Harshbarger and R. Maley (Eds.) *Behavior Analysis and Systems Analysis: An Integrative Approach to Mental Health Programs.* Kalamazoo, Michigan: Behaviordelia, Inc., 1974.

Jones, R. J., & Azrin, N. H. "An experimental application of a social reinforcement approach to the problem of job-finding." *Journal of Applied Behavior Analysis,* 1973, 6, 345-353.

Tharp, R. G., & Wetzel, R. J. *Behavior Modification in the Natural Environment.* New York: Academic Press, 1969.

Donald C. Klein

The relationship of mental health staff and programs to the processes and problems of social change has become a major concern in both mental health manpower training and program development. As the rapidity of social and technological change has accelerated, mental health agencies have become acutely aware of the need to influence broad social-environmental factors in community life. We have, in a relatively short period of time, vastly expanded our perception of what are legitimate mental health concerns.

As this has happened, significant issues have been raised regarding the trained manpower necessary to effect broad aim community programs. In the following chapter[1], Klein[2] addresses some of the conceptual problems involved in this expansion of mental health manpower and programs. He also suggests some of the strategies of action appropriate to these newly created roles, while specifying some guidelines likely to maximize the programmatic effectiveness of change agents in mental health.

In whatever descriptions one might create to sketch highlights of communities in the United States today, the characteristics of rapid change and increased politicization are likely to be prominent. The ecological systems of communities are under considerable stress, and promise to be so for the foreseeable future. It is the difficult problem of training mental health manpower to both preserve the strengths of these natural community systems, and to help effect needed community changes in these same systems, that is and will be critical for the years ahead.

CHAPTER 16

Community Change Agents

INTRODUCTION: A FANTASY

I would like to begin this discussion of community change agentry in the mental health field with a fantasy. Imagine that you direct a comprehensive community mental health center. Your board has directed you to mount a major program of primary prevention. Such a program is to include the following:

1. Maintaining close contact with and seeking to influence decisions of planners and policy makers in government and voluntary social agencies as they affect those aspects of the physical and social environment bearing on the emotional well-being of all citizens.
2. Helping to build and maintain the highest possible degree of community integration (i.e., fostering a sense of community between groups, reducing intergroup paranoia, facilitating good communication both horizontally and vertically, helping to bring to the surface and resolve conflicts, enabling officials and nonofficials alike to respond to citizen concerns before crises have developed and polarization has occurred).
3. Ensuring that adequate educational, child care, health, recreational, law enforcement and other human services are readily available and effectively utilized by all needing them.
4. Making sure that basic supportive and helping services to individuals and families are available as needed round the clock so that developmental problems and the usual crises of living are

dealt with promptly and with a suitable level of caring and support from the community.

5. Providing a broad-gauged program of educational and training opportunities to help citizens develop the psychological insights and skills needed to function more effectively and joyfully as spouses, parents, neighbors, and civic leaders.

Having suppressed the impulse to resign and return to the simple delights of private practice, you consider the requirements of the staff people needed for this enterprise. Apart from specific technical skills (e.g., working with groups, consultation, community organization, education, and training), you come up with the following:

1. a high energy level;
2. broad experience in community activities;
3. expectation of remaining in town for at least ten years, if not a lifetime;
4. ability to work at all hours of night and day, including weekends;
5. ability to mix with people from various strata of the community; and
6. good organizing skills.

You review your present staff members, most of whom were hired because of their professional training for clinical work and consultation with other professionals, and decide that few, if any, have either the savvy or the stamina. You review the budget and realize that only limited amounts of money are available for the new effort. It is not possible to hire a full complement of specialists for the job.

Again having set aside the impulse to resign, your attention is drawn to an article in the *Community Mental Health Journal* describing the use of specially trained lay people to mobilize and implement a job training and employment program for ex-mental patients. Here is the source of manpower you need! You know from experience that some citizens have high energy levels; a few of them on your board are able to drag discussions on long past the limits of your own fatigue level. Who else has as much experience in community activities? Despite a high population turnover in the community, you know that some residents are genuine "locals", living their entire lives in this place of their birth, unlike yourself and most of the center's professional staff members. You don't know about the kinds of working hours they're prepared to keep but surely some of them could be prevailed upon to work swing shifts on crisis lines and the like. The local community college might be a source of night time and weekend person power. Among all those residents there are doubtless some who manage to cross social and ethnic barriers to mix with people from various walks of life. And you know that the community abounds with good organizers; consider the vast number and variety of civic groups that are alive and flourishing! And so you decide that, given proper training and professional supervision, a team of citizen nonprofessionals should be put on the job. Anyway, what choice do you have?

FACTORS FAVORING INVOLVEMENT OF NEW WORKERS

Leaving the fantasy, let us take a look at the reality of nonprofessional involvement in mental health work. The first thing we notice is that there is a growing use of citizens in mental health programs. This use of other than established professionals has been stimulated by the confluence of a number of forces. The first was the forecast of a dire shortage of mental health manpower in the 1960's underscored in Albee's influential study for the Joint Commission (Albee, 1959). The second was the rapid proliferation of mental health agencies stimulated by passage of federal comprehensive community mental health center legislation. This legislation alone made certain that Albee's forecast would come true.

A third force arose, in part, out of the demonstration by Rioch and co-workers that intelligent, mature, college-educated wives and mothers could, if given two years of intensive training, function well as mental health counselors (Rioch, Elkes, and Flint, 1965). The most surprising aspect of this demonstration is that it was needed in the first place. Nonetheless, many established professionals, previously skeptical, were convinced by the results that psychotherapy could be performed competently by others than fully trained psychiatrists, social workers, or psychologists. A fourth force was the mobilization of the so-called war on poverty coupled with the growing realization that mental health programs, like many other human services, were not delivering usable services to ghetto dwellers and other disadvantaged groups. Beginning in the 1960's programs were instituted to train human service aides (Klein, W. L., 1967) and other brands of "indigenous nonprofessionals" most of whom were recruited from the ranks of relatively unschooled slum dwellers (Reiff and Riessman, 1965).

Finally, there was widespread recognition that close working relationships with a wide variety of other resource persons besides established mental health specialists would be needed if mental health programs were to move successfully into either primary prevention or mental health promotion. It was generally acknowledged that, with few exceptions, mental health professionals were themselves laypeople in the educational, organizational, and consultative approaches required for such work. Mental health teams, moreover, often lacked access to and acceptance by those community groups with whom programs of primary prevention and promotion must be involved. It is also true that many mental health professionals are reluctant to invest the long hours, evening and weekend time required to ensure the success of programs designed to alter basic patterns of community life.

Associate Degree Programs in Mental Health

One indicator of the rapidity with which the field has been opening up to the influx of other than established professionals is given by the development of mental health worker training programs by community colleges throughout the country. A national survey recently showed that "the Associate Degree mental health worker movement, which began with five programs in 1966, has experienced a geometric increase in the last six years so that there are now about 150 active programs" (Young, C., True, J., and Packard, M., 1974). Responding to the mental health manpower shortage of the 1960's, community colleges in forty-four states had geared up to turn out approximately 2700 people. The survey projects as a conservative estimate that those programs already instituted, not taking into account other programs that are almost certain to be inaugurated, *will have graduated as many as 17,000 mental health workers by June, 1976.*

Despite vast differences in emphasis and curriculum, most of the two-year Associate Degree programs seek to graduate "generalists" (i.e., "people who are familiar with numerous mental health intervention methods and who are able to provide or arrange for a large number of services to clients and families in a variety of human services settings in cooperation with other professionals.") It is not possible from the survey to determine what proportion of the programs seeks to provide the consulting, training, systems analysis and other skills required by organizational and community change agents. As the report of the survey indicates, the titles of the programs "suggest more heterogeneity in the programs' goals and functions than was expected. Some were closely aligned with traditional mental health professions (e.g., social worker aide), while others had a broad human services orientation."

Academically-based mental health worker training programs represent but a small fraction of the movement of new groups into the mental health field. It is by now widely accepted that "individuals drawn from widely divergent educational backgrounds with only brief or no special training have demonstrated that they can function effectively in mental health roles far above the level of the traditional technician or aide . . . It appears likely that an extraordinarily broad range of human beings can be utilized effectively in mental health settings" (Baker, 1972). Another author, after surveying the use of nonprofessionals as psychotherapeutic agents, comments, "Some studies designate nursery school children as nonprofessionals, while others describe the masters-level person with twenty years of therapeutic experience as a nonprofessional" (Karlsruher, 1973).

Varied Backgrounds and Functions of New Workers

One thing is certain. The mental health field is in a state of flux with respect to the use of other than established professionals. The new workers come in assorted ages, amount of education, experience, socioeconomic background, functions performed, and levels of responsibility. Some are paid employees; others are unpaid staff members or ancillary volunteers. I could find no current figures, but even as far back as 1959 an estimated 43,000 nonprofessionals worked without pay in mental institutions alone, according to a study for the American Psychiatric Association (1959). Add the additional thousands of nonprofessionals serving as board members of mental health agencies and additional numbers of political leaders whose understanding and support are essential. Throw in tens of thousands of community caretakers, such as educators, clergymen, law enforcement officials, and family physicians, not to mention such groups as bartenders and beauticians who are being trained in some places to serve as informal counselors and referral agents (Alcohol and Health Notes, 1972). It becomes apparent that there are enormous numbers and kinds of people engaged in mental health work, or in mustering community support for it, who are not trained in any one of the established mental health disciplines.

Problems of Classification

The range and variety of backgrounds and functions of new workers defy attempts to develop useful systems for classification. An early effort utilized amount of education and came up with three categories of nonprofessionals: specialists, sub-professionals, and aides (Richan, 1961). A more recent scheme put forth by the Mental Health Manpower Commission of Kentucky is based on amount of general education and specialized mental health training, community of origin, and nature of functions to be performed (*Karuna*, 1973). According to the brief report of the Commission's work in *Karuna*, the newsletter of the Center for Human Services Research, the Commission is operating on the assumption that:

> There are at least three identifiable levels of nonprofessionals who can be trained to perform (a) some of the duties and responsibilities traditionally assigned to the professional staff, and (b) other services which are not currently being provided . . . The *first* level is an indigenous member of the community whose function will be to provide various services to the patient and his family. The aide will receive four months of intensive training. The *second* level is an individual with a minimum formal education of a bachelor's degree, preferably with a major in one of the social sciences, who will serve as a community expediter and promote mental health programs among community agencies and organizations. A *third* level of worker is a graduate of an associate of arts program devoted to mental health, case work, psychological testing, group work, record keeping, etc. (*Karuna*, 1973).

Major emphasis in such classification schemes is on the worker's functioning with respect to the primary psychotherapeutic function of the mental health agency. The nonprofessional psychotherapeutic agent usually occupies one of three statuses:

1. His or her functions are ancillary to or supplemental of the functions of the established professionals;
2. The nonprofessional plugs the manpower gap by performing essentially the same functions as the professional, presumably at a somewhat lower level of effectiveness;
3. The nonprofessional works with those clients, usually persons coming from the same background as the worker, who cannot be treated successfully by most professionals.

When the horizon is extended to consider the nonprofessional as change agent (involved in modifying the deployment of the mental health center's resources; establishing outreach projects in the community, facilitating development of new human services delivery patterns, or engaging in fostering the emergence of effective community leadership and problem solving resources) it becomes apparent that there is need for other distinctions that, to my knowledge, have neither been articulated nor tested in the field.

Moving from agency in-house treatment functions to community work in the interests of primary prevention, we immediately come face to face with the fact that some of the most effective mental health agents in the past, beginning with people like Clifford Beers, have been knowledgeable, influential and effective community leaders — nonprofessionals every one. Citizen board members continue to play important parts today in the creation of new services, mobilization of funding sources, and coordination of mental health programs with other agencies in the community (Meyers, et al., 1972). Even local political figures — aware of constituents' personal and family problems — play crucial roles in establishing and maintaining community-based programs and in encouraging their use by those in need (McGee, and Wexler, 1972).

Problems of Nomenclature

It probably makes little sense to call such community figures "nonprofessionals". It makes about as much sense to call the psychiatric director of a mental health center a "nonvolunteer" or "non-community leader". However, the labels we have reflect the frame of reference of those who use them and we have only a few terms from which to choose. One of them — "subprofessional" — can be discarded; it has fallen into disrepute because it so blatantly expresses the conviction that the new worker is beneath the established professional in competence. The term "new professional" is a relatively recent entry. It has the advantage in an egalitarian age of setting aside the sticky matter of differential status and competence; it is content simply to assert that the worker comes with a new background of training and experience. "Paraprofessional" also is an improvement over "subprofessional" since, according to the dictionary I consulted, "para" may mean either "beside" or "beyond" (though unfortunately in medical usage it also refers to that which is faulty or abnormal).

Though the various terms have sometimes been used as if they are interchangeable, I believe that "nonprofessional", "paraprofessional", and "new professional" are by no means synonymous. It is not simply a matter of selecting the least invidious term. Over and above the basic implication common to all three terms that the new colleague is not a member of an established mental health discipline, it is possible to make useful distinctions between the terms and thereby begin to distinguish between the various subgroups of new workers.

I propose restricting the term "new professional" to those, like the two-year Associate Degree graduates, who have received extensive professional training in mental health, however unorthodox that training may appear in comparison with standard career development. Rioch's mental health counselors would be new professionals. New professionals are primarily identified with and committed to the mental health field. They turn for employment to mental health or related institutions. They are dependent on the acceptance of mental health colleagues and on decisions by state civil service bodies to create appropriate slots for them within state supported institutions.

"Paraprofessionals", it is proposed, should be reserved for those professionally trained individuals outside the mental health field (the so-called caretakers or caregivers) who staff the several human services agencies of the community. These individuals do indeed work "beside", and in many cases "beyond", the mental health team when it comes to having a broad impact on the community's emotional well-being. For the most part they are not employees of mental health agencies. Nonetheless, they are essential allies in any effort to develop programs of primary prevention.

Finally, the term "nonprofessional" can be applied to those laypersons who work with mental health programs at various levels — from volunteer aide to agency policy maker, from community action leader to facilitator of an improved interagency referral system. All of them are "indigenous" to some segment of the community, be it neighborhood, ethnic group, or social stratum. Some of them may be specially trained and employed by the mental health center; others are needed as policy leaders, committee members or in other community leadership roles.

For the balance of this paper the three terms are used as defined above.

Problems of Status, Acceptance and Utilization

With the growth of new careers programs in the human services, the problem of nomenclature becomes ever more troublesome. What are the new professional graduates of Associate Degree programs to be called: aides, technicians, assistants? All such terms imply inferior status to those who are aided (i.e., those who are established scientists and professionals as distinct from ordinary technicians). The fact is that the established professionals have gotten there "fustest with the mostest." All additions to the team appear to suffer the semantic and other outrages usually perpetrated on interlopers and newcomers to the tight-knit established communities.

New professionals themselves, of course, are usually aware that they are being subjected to invidious comparisons and put-downs. At a recent graduation ceremony for mental health workers trained at the Hill-West Haven Division of the Connecticut Mental Health Center, a trainee approached the problem as follows:

> Another major change was in getting over my mental process of labeling, of putting people in neat little packages with tags . . . And while reflecting on this in myself, I began to wonder, if I could overcome labeling, why it is that the professionals in mental health find it necessary to continue to label us, refusing to accept us for the skills and knowledge that we have to offer, and why they can't put aside their petty fears of community people taking over their jobs. The mental health field is an expansive field with room for professionals and new professionals without there ever being danger of infringing on anyone else's turf. If only we could work in harmony ourselves, what tremendous output we would have for our clients. (Stuckey, 1973)

As already suggested, the problem is more than semantic. New professionals in mental health, beginning with Rioch's transformed housewives, have not found a wide open job market. Many of them have ended up in human service positions outside the mental health field. Others have found that, without advanced degrees in established disciplines, they top out rapidly and run the risk of ending up as technicians with limited prospects for promotion.

All studies of the utilization of new professionals which I have seen have noted obstacles involving lack of acceptance or insufficient cooperation from senior staff colleagues and supervisors. This was true of Rioch's mental health counselors, of Cowen's mental health aides (Cowen, 1967), and the graduates of nine different two-year Associate Degree programs recently surveyed (Baker, 1972). Of the thirty mental health generalists studied, eighteen reported staff resistance as an obstacle to their fulfillment of role responsibilities. Commenting on the findings, Baker says:

> It is to be expected that anxiety and, to some degree, resentment will be aroused in professionals when they perceive that individuals with considerably less training than that required to attain professional status in their discipline are assuming responsibilities and effectively performing functions previously ascribed only to professionals.

Selection of New Workers

We are only just beginning to address ourselves to the problems of selection of other than established workers for the various roles they are to play. It is hardly likely that a common set of criteria would apply across the board because of the array of functions and settings with which they are involved. Wide differences in interest patterns, value orientations, emotional stability, coping patterns, interpersonal skills, and leadership proclivities are likely to be found when one lumps together such disparate groups as mental health aides working with emotionally-disturbed children and their families, college students manning a hotline program, and neighborhood residents who, following a workshop on community organization, band together for action to reduce crime in the streets in their area.

The little evidence does suggest that selection criteria should, in part at least, be specifically tailored to the particular functions to be performed. An important distinction appears to be that between leadership and therapeutic qualities. The latter include those characteristics usually assumed to be desirable in the mental health field (i.e., warmth and empathy, self-awareness, sensitivity, being aware of one's own personal needs, optimism regarding patients' progress, and the like). The former include such characteristics as ability to handle stress, ability to relate to those in authority, social confidence, comfort in a group, etc.

In one study trainees selected to serve as mental health community project leaders were given the task of organizing and staffing mental health services for particular client populations which utilized community facilities, resources, and volunteer personnel. Examples of actual projects successfully established were a school for retarded children and a twenty-four-hour crisis center (Neleigh, et al., 1971). The trainees initially selected agreed with professional raters that therapist criteria were most important. Later it turned out, however, that "the group who stayed with the job and successfully started projects adjudged themselves as qualified on characteristics which were equally divided between those included as criteria for leadership and those for therapy."

Neleigh and associates noted that nonprofessionals in mental health have represented many different subject populations on the dimensions of intelligence, social position, education and the like. Commenting on their own findings they state:

> The data lead one to speculate that the effective mental health worker may also represent more than one population. Defining the still hypothetical therapeutic personality assumed to be important in a close therapeutic relationship may be an answer for close interaction roles such as psychiatric aides and one-to-one therapists. It may not be an answer for many other roles — coordinators, leaders, educators, developers — which fall within the concept of community mental health (Neleigh, et al., 1971).

The findings of Neleigh, et al. tend to confirm the criteria used by Hallowitz and Riessman (1967) to select neighborhood service center aides. In both cases the most promising criteria represented a mix of therapeutic and leadership qualities. What can be concluded from these findings? It is probably safe to say that the qualities essential for successful one-to-one counseling are not sufficient for the more active roles of program developer, organizational facilitator, or community change agent. It remains to be seen whether the qualities needed for being a therapist tend to be counterproductive for the community change agent.

DIMENSIONS OF CHANGE EFFECTED

Internal to Centers

The advent of new groups of workers in mental health has already resulted in changes affecting the organization of mental health agencies and their service delivery. One of the most far-reaching impacts—usually unintended by those responsible for developing new professional programs — has been on the internal structure of authority, power and policy making in certain large urban mental health programs. New professionals in these settings have politicized the governance and administration. In a few instances they have wrested control from university or hospital-oriented boards; they have mobilized citizen support to ensure client participation in policy decisions, in hiring and firing of directors, and in decisions affecting the deployment of services. In the process of being agents of revolutionary change, such new professionals have alerted the entire mental health field to the importance of re-examing the training and hiring of professionals and new professionals alike with an eye to eliminating institutional, if not individual, racism, and to ensuring citizen participation in the mental health of their communities

(Shaw, and Eagle, 1971). While a few people had been chipping away at such problems for years within training programs and agencies alike, the advent of new professionals gave a needed impetus that in many cases broke through indifference, precedents, and established procedures.

Outreach Programs

For the most part new professionals have been used in more domesticated ways as part of outreach programs, serving as extensions of a center's professional staff into a target community area. The latter has been both geographic (e.g., the neighborhood served by a storefront center), and functional (e.g., the prospective clientele served by a suicide prevention program). In such programs the change agent function of the new worker is ancillary to the direct help-giving role. It has, nonetheless, led to important modifications of human services delivery patterns. Informed by the nature of their clients' problems and their difficulties in securing help from existing agencies, including programs of the mental health center itself, new professionals have sometimes become embroiled in efforts to alter the intake, screening, caregiving, and referral practices of the institutions to which their clients turn for help. Indeed, some mental health programs recognize that outreach programs provide the parent agency with valuable epidemiologic field intelligence about the nature and distribution of problems in the community. In such settings, new professionals may serve as important information sources and stimulators of change within caretaking agencies.

Categories of Outreach

It may be useful to divide outreach programs using new professionals into two categories: the first — and probably more typical kind of effort — involves establishing or extending a program of direct service that becomes part of the continuing programs of the mental health center itself. Examples would be storefront neighborhood mental health stations or neighborhood half-way houses administered by comprehensive mental health centers. Such outreach approaches are being used in community oriented projects on some college campuses.

A recent report of a twenty-four-hour Telephone Counseling Service at Florida State University, manned entirely by nonprofessionally trained students, stresses that the program itself is part of an effort to promote student responsibility-taking and initiative. The Telephone Counseling

Service has stimulated requests for service for a range of problems that "goes beyond the scope of any single agency or any one professional discipline" (Kalafat and Tyler, 1973). One result has been that the counseling center has been encouraged "to promote cooperation and coordination at several levels within the Counseling Center staff and between agencies in the university and community." The Service has also helped the Counseling Center to develop the type of liaison with students which "it has needed in its attempt to enter the university community as a non-traditional, preventively-oriented agency."

The second category comprises those human services initiated and established by the mental health center, either alone or in conjunction with other agencies, that are later separated from the center and enabled to function as independent agencies or be subsumed within a non-mental health institution. An example, already mentioned, was carried out by the mental health service of Las Cruces, New Mexico, using trained nonprofessionals as mental health project leaders. The nonprofessional project leader was trained by the mental health team "to cooperate with the community in planning, promoting and developing a service project. When the project was established, the nonprofessional was to be the project director, responsible for everything from finding and training volunteers to giving service to clients." (Neleigh, et al., 1971). Once established and functioning, responsibility for the projects was turned over to the community.

Influence on External Agencies

We are all familiar with the practice of involving other than established mental health workers in community-oriented mental health programs by means of consultation and education programs. The usual objective has been to help such paraprofessionals perform their caretaking functions with improved attention to their mental health implications. As Caplan (1970) points out in his book on mental health consultation, client or consultee-oriented case consultation provides an entry point into a system which sometimes leads the consultant to become involved in projects of organizational change, program development, and administrative reorganization. Increasing numbers of business and other organizations call upon consultants to assist with internal organizational problems and to help foster programs of continuing organizational renewal. Caplan's description of administrative consultation by the mental health consultant is similar in most respects to the organizational development (OD) approach used by management consultants trained in the applied behavioral sciences.

We discovered early on in the work of the Human Relations Service in Wellesley that it is essential for the mental health team to have at least one ally within each consultee organization in order to work successfully with the consultee system on matters having to do with system change. The ally — sometimes termed the internal change agent — must have access to, and confidence of, key individuals and groups in the system. The internal change agent role is almost always performed by nonprofessionals or paraprofessionals in mental health. To function effectively the internal change agent must be trusted by the mental health consultant and treated as an equal in the collaborative process.

Comprehensive Change Possibilities

Several years ago I pointed out that comprehensive community mental health centers, in part because of the requirements imposed by federal legislation, "are rarely, if at all, committed to any important efforts at environmental modification and community change" (Klein, 1968). I suggested that the mental health field might be in a transitional phase "toward a comprehensive population-centered and preventively-oriented mental health effort." During the transition, concern for the diagnosis and treatment of mental illness was being returned to localities. Today, with the projected withdrawal of massive federal funding for centers and the advent of federal revenue sharing, many comprehensive mental health centers are facing a crisis of survival. The dangers are obvious. Some centers may well go out of business altogether. Others will find themselves restricting services and relying ever more heavily on traditional direct services — both inpatient and outpatient — because they can be supported most readily via patients' fees and payments from Blue Cross-Blue Shield and other insurance carriers.

There is the alternative possibility, however, that some centers may operate from a strong base of local community support to work wholeheartedly on the quality of the community's environment as a physical, social and emotional habitat. To do so will require the active involvement of citizens from various segments of the community and helping them to function as internal change agents for the community's caretaking, planning, coordinating and governing bodies.

COMMUNITY META-FUNCTIONS

Comprehensive community change efforts require major shifts in the deployment of other than established mental health workers as change agents. A quantum leap in both thinking and practice occurs when one leaves the relatively protected framework of the community's human services institutions and becomes interested in the complex matrix of planning, decision making, conflict, collaboration and other basic processes which determine how the social and physical environment affects residents. After reviewing the many different functions ascribed to communities by such sociologists as Sanders (1958), I have concluded that there are three over-arching or meta-functions of community as habitat that are of special concern to the mental health field. They are:

1. Ensuring security and physical safety;
2. Making it possible for people to derive support at times of stress at various points in the life cycle; and
3. Enabling each individual to find one or more anchoring points where selfhood could be established and a sense of significance in the scheme of things developed.

The mental health field, often through the use of paraprofessionals and nonprofessionals, has been making relatively good progress in relation to the first two meta-functions. There are, for example, a growing number of promising examples of work with law enforcement officials, court systems and the like, designed to support the law enforcement process while helping it be more responsive to the emotional needs and life styles of various population groups. Similarly, work with the caretaking institutions of the community has become a keystone of most mental health programs, thus relating the field to the community's institutionalized provisions for helping people manage the usual problems of living.

Fostering Selfhood and Significance

The field has been less certain in pursuit of approaches that might ensure each individual a better chance to find one or more anchoring points where selfhood could be established and a sense of significance in the scheme of things developed. There are three general and mutually complementary pathways that I believe hold considerable promise, each of which involves major commitments of time and energy from others besides established mental health professionals. The first involves the character-building organizations and processes of the community; the second embraces those institutions and temporary systems that provide a supportive context for self-exploration, emotional expression, and interpersonal intimacy; the third entails developing community support systems that facilitate consensual communication between different strata, that make it possible "for individuals and groups to experience interdependence and to encourage leadership efforts in the community," (Klein, 1968) and that fosters, in the terms of Alexander Leighton and his co-workers, *community integration* rather than disintegration (Hughes, Tremblay, Rapaport, and Leighton, 1960). I would like to explore briefly some aspects of each pathway.

Character Building

A characteristic of a rapidly changing society appears to be the lack of stable and enduring character-building institutions or practices that are acceptable to most people. Large numbers of young people today, for example, fall outside the net of established character-building organizations (e.g., scouts, Camp-Fire Girls, 4-H, Y's, local boys' and girls' clubs). Some additional young people become caught up in alternatives, including such varying opportunities as Outward Bound programs and their many imitators, street corner gangs, hot rod, and other special interest groups.

In 1970-71 National Training Laboratories Institute for Applied Behavioral Science conducted an action-research center in the new city of Columbia, Maryland known as the Community Research and Action Laboratory. Asked to study the needs of young people and the organizations serving them, a Laboratory team discovered a great variety of recreational and other programs offered to teenagers by the adult community and a paucity of opportunities for young people to engage in significant ways in the life of the community itself. The adolescent years provide young people with few character-building opportunities related to building a sense of significance in the larger community of which they are, all too often, a marginal and disturbing part.

Mental health programs could help to stimulate and initiate projects that would involve young people significantly in communal affairs. They could help train the nonprofessionals and paraprofessionals working with young people and support their efforts to involve teenagers more significantly in planning for their lives. One approach might be to engage the services of young people themselves to help determine what life in the community is like for various subgroups among them.

Supported by an NIMH grant, we are beginning such a process this year in Columbia. The project involves hiring a team of 15 to 18-year-olds to carry out a study over a one-year period. Some of these nonprofessionals will serve as participant-observers, keeping detailed logs of their activities, contacts, community involvements, and other activities. Simultaneously we are establishing a steering committee of both young people and adults. They will be divided into at least three work groups, one to facilitate youth involvement in the project, the second to assist with planning for data collection and analysis, the third to work on feedback of findings in ways that hopefully will facilitate needed problem solving and change. This project, if successful, may serve as a model for a relatively inexpensive way for mental health centers to use young people as nonprofessionals in a continuing or intermittent process of diagnosing youth needs, setting priorities, and stimulating appropriate community action.

Supportive Systems for Self-Exploration

The second pathway for fostering selfhood and significance involves social institutions and temporary systems that provide support for self-exploration, emotional expression, and interpersonal intimacy. Here too, the official mental health organizations have generally left the field in the hands of others (e.g., growth centers, religiously-inspired programs of self-exploration, and the many self-appointed encounter group leaders who have proliferated in many communities and on college campuses). Questions have been raised about the possible damage to participants that can be inflicted by poorly led activities of this sort; others have questioned whether encounter groups generally induce lasting behavior and personality improvement. Meanwhile, however, the groups continue and participants agree that they have gained a new sense of their own worth and potential from the self-affirmation afforded by such "intimacy training".

There are now many competent and responsible paraprofessionals in the humanistic psychology movement. They represent a valuable set of resources for fostering human potential, increasing parent effectiveness, and promoting mental health generally. Their numbers could be augmented within any locality should the mental health center choose to invest in the further development and support of such workers.

It is not a far cry from the temporary sense of community to be found in most encounter groups and training laboratories to the growing numbers of attempts to foster personal fulfillment and a richer life style in communal experiments and lasting intentional communities. Recently I spent a year travelling around the United States and in the process contacted several communal groups, individuals who were looking into such experiences for themselves, and two intentional communities. I found a generally high level of interest in what psychology and the mental health professions generally could offer such experiments. Certainly the concerns with which such groups must deal are not alien to the mental health field. They include issues having to do with interpersonal pair bonds, jealousy, sexual experimentation, money management, authority and power, and child rearing. A few urban communes are reported to be making use of the periodic services of encounter group leaders and mental health consultants. My own work with the Twin Oaks Community in Virginia has taken the tack of helping prepare some of its members to serve as coordinators and facilitators of feedback sessions and other community gatherings designed to work on community maintenance and tension reduction.

Developing Linkage and Problem Solving Systems

The third approach to fostering selfhood and significance entails developing support for intergroup contact, communication, collaboration and problem solving. A few such projects have been initiated — most of them by local community development agencies, church-supported urban centers, and university-based programs for the application of behavioral sciences to community affairs. However, one of the most extensively developed examples was inaugurated by the Prairie View Mental Health Center in Harvey County, Kansas (Raber, 1972).

The Center had a tradition of using Maxwell Jones' therapeutic community concepts internally in an effort to create an atmosphere of openness, maximum patient responsibility and a milieu of trust and mutual interdependence within the treatment programs of the institution. As it became increasingly involved with local citizens and their organizations, the staff set about developing a support system in the community that would foster interdependence among groups and encourage the emergence of community leadership. It did so by helping launch an independent organization of citizens known as Leadership, Inc. (LINC). The purpose of the organization was "to improve the support system, emotional climate, and the quality of living" in the county. Using laboratory training methods, such as those developed over the years by persons associated with National Training Laboratories (NTL), groups of community residents have come together for experience-based learning both about themselves and about community problem solving.

An effort similar to LINC was initiated in 1967 in Middletown, Connecticut, as the result of an alliance between a Community Development Corporation established by Wesleyan University and Dr. Herbert Shepard, an applied behavioral scientist in the field of organizational behavior at Yale University. To paraphrase from the group's original announcement of its program, Middletown Future is a program designed to enlist the creativity of the people of Middletown in developing a community that is better able to meet the needs of all of its citizens, and that can lead to greater self-fulfillment and a greater understanding of our fellow citizens for each of us.

Under a one-year grant from the State's Department of Community Affairs nineteen workshops were conducted over a two-year period. Most of the programs brought together maximum mixes of citizens from different backgrounds. In addition, there were "specialty workshops" for business executives, teen-agers, police, black community leaders, clergy, realtors, school administrators, and others.

Both LINC and Middletown Future drew on the experiences of applied behavioral scientists who have experimented with adapting T-group and laboratory training approaches generally to organizational change and community development efforts. In one such precursor I was hired by the paper mill management of a small one-industry community to work on the problem of citizen apathy and unconcern about the physical and social well-being of the town (Klein, 1965). I undertook the project initially as a staff consultant from the mental health program of the Massachusetts General Hospital. So far as I know, it is the first such attempt by a mental health program to deploy methods from the small group training field in a community development project.

A three-step program was carried out:

1. Fact-finding efforts with twelve community leaders, the results of which were shared with the interviewees, with top management of the company, and with an advisory committee of the mill's middle-management, most of whom were directly con-

nected with the operations of the mill itself and all of whom were community residents;
2. Periodic meetings with an ongoing group of community leaders to explore community needs, establish priorities, provide a mutual support system for individuals' efforts at community improvement, and test the feasibility of a long-range community development program; and
3. Training of teams of leaders as facilitators of community change so as to maintain a change agent capability in the town once the outside consultants were no longer available.

SOME ISSUES ABOUT CHANGE STRATEGIES

In the above three examples there are certain underlying issues which warrant special consideration:

1. These early attempts place a major reliance on experience-based learning approaches more or less derived from T-group methodologies and sensitivity training. They are relatively heavily skewed towards a process (as vs. project) orientation, emphasizing how people relate to one another, how well they communicate, support one another, and handle conflicts rather than focusing primary attention on tasks worked on and results achieved. Process approaches are highly satisfying to many persons. They encourage a climate of trust and enable participants to develop mutual support patterns. They lend themselves, as the Prairie View team discovered, to the adaptation of the therapeutic community approach to a geographic politically-defined locale.

However, there are some community leaders who mistrust the process approach. They are geared to working on "real" community problems, such as housing, racism, economic oppression, environmental pollution, integration, and maintenance of neighborhood schools. They are not inwardly directed and are not comfortable with discussion of individual feelings and relationships. The process approach leaves them indifferent, bored, hostile, sometimes even scared.

2. Closely related to the first issue is the tendency on the part of practitioners and devotees alike to equate the process approach with a collaborative style and value orientation. Community problems are viewed as being capable of resolution by people of good will provided they possess the interpersonal skills that behavioral science can provide

and if they are helped by means of suitable change agent procedures to function in a climate of mutual trust and support. The collaborative approach is profoundly mistrusted by many community organizers accustomed to working with poor people and disadvantaged minorities. In the hands of the establishment, they maintain, the approach represents co-optation of the opposition. They point out that genuine collaboration can occur only among those possessing roughly equal power in relation to one another. Until equality has been established, the collaborative process has the unfortunate effect of humanizing the enemy and thereby weakening the protesters' resolve to struggle against and overcome the opposition.

Saul Alinsky was the best known exponent of the conflict approach to community action. From a radical labor union background, he mistrusted a value orientation that sought out commonalities among groups which, he believed, were in fundamental opposition to one another. His emphasis was on mobilizing solidarity among the underdogs by scoring achievable victories against the topdogs. Though Alinsky seems, on the surface, to be a results rather than a process-oriented change agent, I believe that within his own value orientation he was the epitome of a process-oriented consultant. For him community action was the best process designed to educate the oppressed, to give them skills as well as hope, to motivate them to band together, and to teach them how to take control of their destinies.

A Developmental Approach to Community Change

In point of fact, there are few, if any, change agents working from an established mental health base who are in a position to apply a pure conflict model. There are simply too many financial, and other strings attached to mental health agencies by funding and sanctioning sources. Financial and other support would be seriously jeapordized by any sustained effort to confront the establishment or redistribute power relationships within the community. Other means must be found, therefore, to intervene in minimally threatening fashion to improve community life if mental health agencies are to fulfill their basic missions of primary prevention and health promotion. A project in which I am currently involved, known as the Developmental Approach to Community Change, represents an effort to refine such an approach and disseminate it among mental health and other agencies engaged in community development activities involving nonprofessionals as part of the change process.

Based in Richmond, Virginia, the project takes its point of departure from the basic premise that the overall physical and psychological well-being of a community's population is, in part, a function of the community's human service agencies response to the needs of residents. The decision was made to focus on improving the delivery and utilization of three kinds of human services as a vehicle for building the intergroup trust, interdependence, and leadership skills needed in the community. The methods used include:

1. Experience-based training with teams of workers and administrators from three city agencies — police, welfare, and recreation — and with counterpart teams of citizens from three target neighborhoods;
2. Consultation and third-party intervention techniques to help bridge gaps between deliverers and users of services; and
3. Simulation methods that enable practitioners and users, working together, to plan and implement effective solutions having to do with specific problems of service delivery.

The Project makes use of three levels of change agents:

1. A national resource team of behavioral scientists provided by NTL;
2. A continuing local group of professional and paraprofessional consultants and trainers working out of the organizational base of TRUST, Inc., a nonprofit community development agency located in Richmond itself; and
3. Nonprofessional training assistants and consultants recruited from the neighborhoods themselves and trained to serve as volunteer facilitators of community problem-solving within their own locales.

An important feature is the attempt to forge an effective alliance of a national nonprofit organization (NTL) with a local organization for the delivery of behavioral science skills (TRUST, Inc.). The original assumption — based on OD experience in single institutions — was that both "inside" and "outside" change agents would be needed to ensure, on the one hand, objectivity, and on the other, the continuity of concern and understanding that only an in-house change agent team can furnish. It soon became clear that in the macrosystem or "system of systems" that is the community, it is essential that equally strong partnerships and working agreements be established between the in-community change team from TRUST and the several autonomous organizations and neighborhood groups involved.

Memoranda of Agreement (MOA)

An important working tool has become the process of working out written memoranda of agreement which define, insofar as possible, how each pair of groups in the project will work together, how disagreements will be handled, and what assurances are needed regarding confidentiality, absence of reprisal, and the like. The MOA approach takes into account the complexities of the macrosystem that is the community; it accepts the existence of differences in ideologies, vested interests, sources of support and sanction, and it assumes that continued commitment to the effort can occur only so long as negotiations to maintain suitable conditions for involvement can be maintained. The process is analagous to the development of sanctions which must be accomplished by a change agent working with a large organization at several levels. It is different from the sanctioning process in that the change agent takes the additional step of facilitating working agreements among the several groups involved.

Integrating Project and Process

Equally important is the program's commitment to work out a satisfactory balance between the project and process emphases discussed above. The first step was to secure the interest and endorsement of the City Manager and his immediate staff. The next step was to enlist the cooperation of three city agencies and to develop a task force in each whose mission was the improvement of service delivery in a specific inner city neighborhood. Once possible neighborhoods were identified, it was the job of the project's change agent team to determine whether a citizen task force in each neighborhood could be recruited. Only after each agency and neighborhood task force had come together often enough to develop some cohesiveness and establish a list of problem priorities and concerns were joint meetings of agency and neighborhood task force pairs held.

It is the task of each pair to select a workable problem which, if solved or at least ameliorated during the next year or so, would have a good chance of improving service delivery by the agency in the neighborhood. As task force pairs meet together it is the task of the change agent team to serve as third-party resources to help design meetings, bring hidden agenda to the surface, facilitate communication, and minimize destructive polarization.

Data-Based Intervention

The project is applying to the macrosystem field a finding from organizational development studies to the effect that data-based interventions, which provide new information about the client system for the latter to use in its efforts at improvement, are superior to interventions based solely on the facilitation of interpersonal or intergroup communication, decision making, or other process skills. Nonprofessionals from the community, trained by and working under the direction of a sociologist, conducted extensive stratified sample interviews with neighborhood residents, focusing on kinds of services being delivered and resident satisfaction with them. These data were fed back to neighborhood task forces and used by them in formulating problem areas and priorities to be discussed with the agency group. Similarly, project staff interviewed each member of the agency team about his or her role and functions, job satisfactions, dissatisfactions, and the like. Results of the interviews were fed into the working sessions of each agency task force to be used in the determination of their priorities and in the formulation of their reports to their neighborhood counterparts.

We are now halfway through the three-year NIMH grant period for the action research and it is too early to gauge results. We have just entered into the recruitment and training of the first group of neighborhood training assistants from each target area. To begin with, these nonprofessional facilitators will be working with the neighborhood task forces and other residents to help them develop the self-confidence, communication skills, and problem solving techniques needed "to deal with city hall". Our intention is to break through the frustration and apathy which makes it so difficult to mobilize and communicate citizen concern in ways that care-giving professionals and public officials can understand and accept. We doubt that our neighborhood task force members and nonprofessional training assistants will go *fight* city hall; we expect that they will *influence* city agencies and be a force for improvement of the quality of life in Richmond.

CHANGE FUNCTIONS AND SETTINGS: A CLASSIFICATION SCHEME

Functions

Mental health workers have the potential of performing a number of different community change functions. At this point I would like to suggest a classification schema that juxtaposes functions and the settings in which they are performed. The change agent functions are: action leader; change manager; and facilitator.

The *action leader* provides the sustained drive and direction required to maintain a change process until a problem has been resolved or some action goal achieved. The action leader is committed to a particular solution or set of outcomes.

The *change manager* performs a coordinative function, enabling those most directly involved in a change process to pool their energies, resolve differences, and work together to diagnose needs and determine a course of action. The change manager is usually, though not invariably, a person functioning within the system that is targeted for change.

The *facilitator* is the one who focuses attention on orderly processes of problem solving. In an earlier discussion of this role I said, "They emphasize commonality of interests and community identifications on the part of conflicting factions. They play the roles of mediators, fact finders, facilitators of communication and reality testers" (Klein, 1968).

Using the community mental health center as the point of reference, any of the above three functions can be performed in three kinds of situations: intra-mural; extra-mural; and external.

By *intra-mural* I mean a situation within the mental health center itself having to do with its own program development or internal organizational arrangements and procedures.

By *extra-mural* I mean those situations in which the mental health center is engaged in outreach efforts to extend its resources out into the community. A storefront center, twenty-four-hour drug hotline service, suicide prevention center, or any other external program for which the mental health agency maintains administrative responsibility would be classified as "extra-mural". Extra-mural programs also include mental health consultation services to outside agencies and caretakers.

By *external* I mean those situations or settings for which the center has no administrative or direct program responsibility. Thus, while the mental health consultation service to the public school system would be "extra-mural", the public school system itself would be "external".

By arranging functions and settings in a 3 x 3 table (Table I) it becomes apparent that, theoretically at least, mental health workers and their citizen allies can perform at least nine change agent roles in the community.

TABLE I

Classification of Community Change Operations By Functions and Situations with Examples

FUNCTIONS	SITUATIONS		
	Intra-Mural	Extra-Mural	External
Action Leader	1. nonprofessional mental health aide	4. paraprofessional clergyman	7. new professional mental health consultant
Change Manager	2. new professional project coordinator	5. nonprofessional mental health aide	8. paraprofessional professor of education
Facilitator	3. professional staff consultant	6. university adult educator	9. community development specialist

Some Examples

It may be helpful to examine how all three functions might be performed in each of the settings. The numbers in the following examples refer to cells in Table I.

1. An action leader from within the mental health center would be exemplified by the nonprofessional mental health aide at the storefront center who notes that insufficient involvement of neighborhood residents on the agency's board makes it impossible for residents to trust the agency and make adequate use of its services. The nonprofessional organizes a group of other nonprofessionals and sympathetic professional staff members to put pressure on center administration. It is doubtful that citizens will be hired by agencies for the express purpose of initiating or fomenting internal changes in agency structure or program; it is nonetheless likely that involvement of new kinds of workers from the community will lead to such intra-mural change agentry.

2. An intra-mural change manager would be someone designated by center administration to take the coordinative leadership required to design and install a needed change. A new professional mental health associate degree graduate might be assigned to coordinate a joint staff-board task force charged with the job of designing a suitable way for involving more neighborhood residents in policy decisions.

3. The facilitator from within the mental health center would serve as a consultant to the change process and might work primarily with the task force coordinator to help the latter design and implement the procedures to build a cohesive task force and help it make the best use of its resources.

4. Turning next to extra-mural situations, the action leader might be exemplified by a paraprofessional clergyman who is convinced that the mental health center should be providing a telephone hotline service on a twenty-four-hour basis to help those with suicidal impulses. Though not employed directly by the center, the clergyman has been closely involved with mental health work, has referred parishioners to the clinical service, and has used center professional staff as consultants to him. At this point, the clergyman takes the initiative to mobilize both the community pressure and support needed to bring about a change in center priorities. The decision is made to design and install a hotline service to be coordinated and staffed by paraprofessionals and nonprofessionals from the community.

5. The change manager might be an experienced nonprofessional mental health aide with prior experience and training in crisis management. The nonprofessional is also a capable trainer, able to use role playing and other techniques to help relatively unskilled persons develop the capabilities required for hotline work. The change manager organizes a task force of interested volunteers which contacts local clergymen, service groups, and colleges to recruit workers; designs and implements the training program with the help of selected professional staff from the center; installs the service; and works with the local mental health association and other civic groups to market the new service.

6. The facilitator in this instance might be an adult educator from a nearby university who works with the change manager and hotline team during the first year of the operation. This paraprofessional's efforts are directed towards helping the team manage its anxieties, identify its learning needs, and develop opportunities for further skill development.

7. Turning next to external situations, the action leader from the mental health agency might be exemplified by a new professional consultant to pre-school services in the community who is authorized by center administration to take whatever steps are needed to form a Community Coordinated Child Care Committee (4C). The worker interviews child care leaders and parents, gathers statistics on the number of children requiring day care services in the community, identifies interested local and state agencies, and ultimately convenes a core group of individuals committed to the task of building an effective 4C effort.

8. Once the 4C group has been convened the action leader from the mental health center would be well advised to transfer the impetus for the effort to a change manager from the community who is not employed by the center itself. Such a person might be the paraprofessional professor of early childhood education at the local community college who has the energy, time, and child development know-how needed for the job, and who has the confidence of the key groups involved.

9. The facilitator might be a community development specialist from the local antipoverty agency whose task would be to help the fledgling 4C group navigate its way through the many potential pitfalls of conflicting or competing interests inherent in such an effort to build a community coalition. The facilitator is an ally of the mental health center which, through its training programs on community problem solving for community leaders, has helped the specialist learn new skills and form important linkages with government officials and other civic leaders.

Redefining Change Agentry for Macrosystem Interventions

The above examples, though hypothetical, are intended to underscore the vast variety of resources in the community which can be drawn upon by the mental health center in its efforts to promote needed community change. They also try to indicate the importance of building, *over a period of time*, an investment in an array of change agents in all three situations who are committed to mental health objectives and are capable of performing action leader, management and facilitative functions as needed.

It should be acknowledged that the definition of change agent in this paper differs from that which is ordinarily presented in the literature. In their classic work in the field, Lippitt, Watson and Westley (1958) viewed the change agent as a facilitator of change brought into a client-system from the outside. Mental health consultants, applied researchers, psychotherapists, organizational development training consultants are all examples of generally accepted change agents imported into a client-system from the outside.

In the intervening years the concept of change agent has been extended to include internal facilitators within a client-system as well as those brought in from the outside. Bennis, Benne and Chin (1961) hold, for example, that "client-systems contain the potential resources for creating their own planned change programs under certain conditions; they have inside resources, staff persons, applied researchers, and administrators who can act and do act as successful change agents." These authors contend further that a "client-system must build into its own structures a vigorous change-agent function, in order for it to adapt to a continually changing environment."

In this paper we are considering not a single client-system but a complex matrix of inter-systemic relationships within a macrosystem, all of which are implicated in any comprehensive effort to develop effective preventive mental health programs. The mental health center itself, to add to the complexity, is part of the client-system matrix in that it offers direct human services and thereby contributes to the community's capacity for meeting the needs of its citizens.

Need for Distributed Change Capability

To bring about comprehensive community changes for improved mental health, the capability for initiating, promoting, managing and facilitating change must be well distributed through the various settings, including the mental health center itself. It is not enough for the center to deploy change facilitators as consultants to various external systems. There must also be the intra- and extra-mural capability for taking direct action for change and for managing change efforts once they are underway. The same holds true for the external settings. To paraphrase Bennis, Benne and Chin (1961), the community must have built into its own structures a vigorous set of change-agent functions, in order for it to adapt to a continually changing environment. The mental health center can play a leading part in building such a capability by establishing linkages with and among those individuals and groups capable of working collaboratively on community improvement. Its staff can play an important role, as does the Prairie View mental health team, in building a climate for change and for fostering the skills needed to make the best use of emerging community leadership.

A definite necessity will be involvement with the center, of citizens capable of performing the several change functions.

FANTASY REVISITED

Returning once again to the fantasy with which this presentation began — the mental health center director who has decided to use citizens as the basis for a program of primary prevention — it should be clear that any such coordinated attempt to mount a comprehensive program remains in the realm of fantasy. What we have instead is a variety of efforts to use in experimenting with the use of nonprofessionals, paraprofessionals, and new professionals as change agents.

When I originally confronted my own fantasy as I began writing this paper, it reached such nightmare proportions that I almost resigned from the writing task itself. What a herculean task it would be to even explicate an experiential and conceptual base for deploying new workers in community change efforts! I am pleased that, like the fantisized director of the center, I did not resign. I have emerged at this point in the exploration of the topic impressed with the richness of possibilities for using other than established mental health workers as change agents. Isolated examples exist and we have the beginnings of a conceptual framework.

The past decade has seen the rapid emergence of professionally staffed community-based mental health centers, a development which seemed hardly possible when, in 1953, I assumed the Executive Directorship of the pilot mental health program in Wellesley, Massachusetts. We have since learned how to deploy resources for early case finding, crisis intervention, other forms of secondary prevention and rehabilitation. I think we are now hovering on the brink of extensive programs in primary prevention and promotion of emotional well-being. Such programs will, I predict, be built on a more solid base to the extent that they will make extensive and differentiated use of new professionals and other competent allies drawn from the community itself.

FOOTNOTES

[1] Prepared for the West Virginia Conference on "Issues and New Developments in the Delivery of Mental Health Services," May 6 — 9, 1973.

[2]Program Director, Center for Macrosystem Change, National Training Laboratories for Applied Behavioral Science, Washington, D. C.

REFERENCES

Albee, G. W. *Mental Health Manpower Trends.* New York: Basic Books, 1959.

Alcohol and Health Notes, Vol. 1, No. 1, December, 1972, Rockville, Md.

American Psychiatric Association, *The Volunteer and the Psychiatric Patient*, Washington, D.C.: APA, 1959.

Baker, E. J. "The mental health associate: A new approach in mental health." *Community Mental Health Journal*, 8, No. 4, 1972, 281-291.

Bennis, W., Benne, K. & Chin, R. *The Planning of Change.* New York: Holt, Rinehart, and Winston, 1961.

Caplan, G. *Theory and Practice of Mental Health Consultation.* New York: Basic Books, 1970.

Cowen, E. L. "Emergent approaches to mental health problems." In E. L. Cowen, C. A. Gardner, and M. Zax (Eds.), *Emergent Approaches to Mental Health Problems.* New York: Appleton-Century-Crofts, 1967.

Hallowitz, E. & Riessman, F. "The role of the indigenous nonprofessional in a community mental health neighborhood service center program." *American Journal of Orthopsychiatry*, 37, 1967, 776-778.

Hughes, C., Tremblay, M., Rapaport, R. & Leighton, A. *People of Cove and Woodlot.* New York: Basic Books, 1960.

Kalafat, J. & Tyler, M. "The community approach: Programs and implications for a campus mental health agency." *Professional Psychology*, 4, No. 1, 1973, 43-49.

Karlsruher, A. E. "The nonprofessional as a psychotherapeutic agent: A review of the empirical evidence pertaining to his effectiveness." University of Manitoba, unpublished paper, 1973.

Karuna. Publication of the Center for Human Services Research, The Johns Hopkins University, Department of Psychiatry and Behavioral Sciences, Issue No. 6, March, 1973.

Klein, D. *Community Dynamics and Mental Health.* New York: Wiley, 1968.

Klein, D. "Sensitivity training and community development." In E. Schein and W. Bennis (Eds.), *Personal and Organizational Change Through Group Methods: The Laboratory Approach.* New York: Wiley, 1965.

Klein, W. L. "The training of human service aides." In E. L. Cowen, E. A. Gardner, and M. Zax (Eds.), *Emergent Approaches to Mental Health Problems.* New York: Appleton-Century-Crofts, 1967.

Lippitt, R., Watson, J. & Westley, B. *Planned Change.* New York: Harcourt, Brace and Company, 1958.

McGee, T. F., & Wexler, S. "The evolution of municiaplly-operated, community-based mental health services." *Community Mental Health Journal*, 8, No. 4, 1972, 303-312.

Meyers, W. R. et al. "Methods of measuring citizen board accomplishment in mental health and retardation." *Community Mental Health Journal,* 8, No. 4, 1972, 313-320.

Neleigh, J. R. et al. "Training nonprofessional community project leaders." Monograph Series No. 6, *Community Mental Health Journal,* 1971.

Raber, M. "Leadership, Inc. (LINC): A new model for community change." Prairie View Mental Health Center, Newton, Kansas, unpublished, 1972.

Reiff, R. & Riessman, F. "The indigenous nonprofessional." Monograph Series, No. 1, *Community Mental Health Journal,* New York: Behavioral Publications, 1965.

Richan, W. C. "A theoretical scheme for determining roles of professional and nonprofessional personnel. *Social Work,* 6, 1961, 22-28.

Rioch, M., Elkes, C., & Flint, A. *Pilot Project in Training Mental Health Counselors.* Publication No. 1254, U.S. Public Health Service, Bethesda, Md., 1965.

Sanders, I. *The Community: An Introduction to a Social System.* New York: Ronald Press, 1958.

Shaw, R. & Eagle, C. J. "Programmed failure: The Lincoln Hospital story." *Community Mental Health Journal,* 7, No. 4, 1971, 255-263.

Stuckey, G. L. "Challenges in mental health for new professionals." *American Journal of Orthopsychiatry,* 43, No. 1, 1973, 164-167.

Young, C. E., True, J. E., & Packard, M. E. "A national survey of associate degree mental health programs." *Community Mental Health Journal,* January, 1974.

Conclusion

Roger F. Maley

Dwight Harshbarger

CHAPTER 17

The Integration of Behavior Analysis and Systems Analysis: A Look at the Future?

This book has attempted to provide a framework for the integration of two different, yet complementary, approaches to the solution of human problems. We have seen how proponents of both behavior analysis and systems analysis have developed their areas of expertise and commitments to professional action. Along the way, we have also encountered countless examples of treatment fragmentation, stupidity, poor planning, and generally ineffective approaches to the overall delivery of quality mental health services. As professionals who attempt to rectify some of these difficult problems, our limited past successes clearly indicate that we must try new ways of organizing human services, as well as developing more sophisticated technologies of behavior change.

Although the histories of the behavior analysis and the systems analysis movements are quite separate, they contain many common elements. For example, both groups have been on the fringes of academic psychology and have shared many common foes. Their different language systems have evolved to separate in-group from out-group, as well as to more accurately describe certain behavioral phenomena at different levels of analysis. Although language is a problem, our experiences at the conference which led to this book indicated that with just a little bit of effort the two groups could communicate very well with one another. They share a common interest in such areas as accountability, objective outcomes, the design of rational systems of treatment, and the use of paraprofessional personnel. However, even at the conference, the "turf problem" reared its ugly head from time to time reminding us that we must be vigilant in reducing this problem within the field of psychology and within our dealings with other mental health disciplines.

At a more abstract level, behavior analysis and systems analysis seem to have proceeded along different paths towards common goals. While the systems analyst might quip that this merely demonstrates the principle of equifinality, the behavior analyst is likely to wryly observe that the evidence that common goals are being pursued is not yet convincing. These paths might be described as differing in general approaches to problem-solving which stem from the use of inductive-empirical approaches in behavior analysis, and deductive-conceptual approaches in systems analysis.

In the real world, in, for example, a particular mental health organization, the two groups' problem-solving behavior probably proceeds in the following manner: The systems analyst is likely to map out a conceptual model of the organization, its internal and external resource exchanges, feedback loops, and its social policies establishing the flow of rewards in the organizational environment. He will concern himself with aggregate data on programs and staff/client behavior, and should it be desirable to change the organization, he is likely to choose points of intervention that can be viewed as molar elements of the system; for example, changing policy to retard the acceptance of clients or to accelerate client discharge by the organization. The systems analyst is likely to proceed using a reductionist model, focusing on the behavior of particular individuals after having examined aggregate group or sub-system data.

The behavior analyst, on the other hand, is likely to begin problem-solving on the reverse side of the organizational system. His first focus is likely to be the behavior of individual staff and clients, and he will probably proceed inductively and additively to construct sub-system and, later, large system operations based on the examination of aggregate data. His conceptual map of the organization is built from the bottom up, rather than from the top down. Should he choose to develop organizational change efforts, the behavior analyst is likely to first focus on attempting to alter the contingencies affecting individual staff and/or client behavior. He will proceed upward through the organization to alter group or sub-system contingencies to the extent that it is necessary to bring about the desired behavior change. Given the nature of the beasts called human service delivery systems, it is highly probable that before reliable individual and sub-system behavior occurs, the behavior analyst will find himself concerned with broad organizational policy as a contingency management problem, and the systems analyst will find himself concerned with individual behavior as a derivative of organizational contingency management, or social policy.

The point here is not merely that these two groups of problem solvers will unavoidably be concerned with similar problems; rather, that from the outset they are dealing with different aspects of the same problems. The sooner this is realized, the greater the likelihood that their efforts will be complementary and mutually supportive. Effective behavior change strategies in organizational sub-systems are doomed without the appropriate changes in social policy. Effective social policy rests upon its resulting in more adaptive, rewarding, individual behavior.

At the conference leading to this volume there was a growing awareness by behavior and systems analysts that, first, these common concerns were increasingly obvious; second, that these two groups really did share these common concerns; third, that to be effective change agents these groups needed each other's skills; and fourth, that the potential for integrating the languages of the two groups was relatively high. Systems analysts found their conceptual models more workable through the integration of the functional relationships and specifiable outcomes generated by the language of behavior analysis. Behavior analysts found their problem-solving strategies enriched and expanded through the integration of the larger, more encompassing conceptual models of systems analysis. It is our hope that human service delivery systems will be better because of it.

ISSUES FACING BEHAVIOR ANALYSTS AND SYSTEMS ANALYSTS IN MENTAL HEALTH DELIVERY SYSTEMS

What Problems Are We Trying to Solve?

If the mental health professions are really committed to trying to solve "problems in living" then we must begin to actively intervene in problems resulting from poverty, unemployment, changes in the nuclear family, and social oppression. However, since we do not have the necessary professional manpower to adequately meet the demands of individuals who need or desire help, how can we possibly move into new areas of professional activity? Even if we do not totally accept this broad mandate, the problems of limited resources and ineffectiveness continue to plague us. Something, perhaps something fairly radical, needs to be tried.

The mental health movement, in general, has not dealt effectively with many of the problems which have most concerned society. Many of the failures were due to omissions and to the fact that we promised ourselves and society too much. Unfortunately, we believed the rhetoric which we helped generate in the 1960's, and the larger society, some members of which also believed us, has become acutely aware of our failures. For example, our efforts in the field of delinquency prevention do not seem to have reduced crime; the plights of the mentally retarded and the chronic mental patient have not been dramatically improved; and the cost to society of the multi-problem family, which is impervious to intervention, has not been significantly lessened. The social movement called community mental health and its spiritual sibling, the war on poverty, have not been very effective in achieving their stated goals.

Our students are aware of our "credibility gap" and sense the need for renewed commitments to public stewardship through systems of professional accountability. Because most mental health systems have a captive market with little or no competition to provide treatment alternatives, the systems have not been subject to independent, outside reviews which might promote competency and responsiveness. In addition, the leadership in mental health frequently has been exercised by mental health professionals who have refused to accept public criticisms and demands to explain failures. Such a situation cries out for accountability systems which hold professionals responsible for the results of their therapeutic endeavors and intervention efforts, and reward them according to the outcomes they produce.

How Will We Evaluate Our Effectiveness?

Hopefully, the days are over when mental health or any other group of professionals can justify their activities merely by stating that they are engaged in a process which they "believe" is helpful. Today, it would seem, we need to take a hard look at the objective outcomes of our professional endeavors. This new focus upon objective outcomes may, in some ways, be premature; yet, it seems unlikely that we will be able to make significant improvements in delivery systems until this kind of commitment has been made. If the mental health professions are to have wide-spread public support, they can no longer afford to turn out practitioners who are not vitally interested in evaluating their professional effectiveness. The time has come to take seriously our privileged status as societal change agents and therapists and begin to make our operations more effective.

Many young people are convinced that the mental health professions are dedicated to a grand societal rip-off. These feelings have arisen, at least partially, in response to the demonstrated lack of effectiveness in practically every area of professional involvement. Although this is probably a harsher conclusion than the professions deserve, it is not one that is totally irrational. To argue that our mission of changing complex human behavior is just too complicated and too difficult to be effectively evaluated, is begging the issue, and amounts to a colossal evasion of our current difficulties.

Thus, when we design new systems of accountability, we must insist that they be controlled, at least partially, by people outside of the mental health professions. It's not the case that the professions are dishonest. Rather, external validation is necessary to keep on track in therapeutic efforts and honest in the evaluation of these efforts. Accountability has come to mean a lot of things to a lot of different people, but it must ultimately lead to the types of evaluation that can be consensually validated apart from the jingoism of professional dialogue.

How Will We Organize Our Efforts?

The broad scope changes that the mental health professions need to make are fairly obvious. As has been echoed by many writers in this book, mental health services are far too fragmented and are, frequently, organized for the benefit of professionals rather than clients. For too long now, we have focused on remediation and given only lip service to preventive services. In addition, we have not developed the type of all-encompassing delivery systems which would provide maximum treatment continuity and overall program success. It may well be that we will need to develop new agencies and organizations in order to deal with society's problems in controlling deviance, and to provide the types of help needed by most Americans.

Our old conceptual systems, based mostly on psychodynamic approaches to problems, have not proven effective bases for intervention systems while other models, such as the behavioral model and the public health model, have shown considerable promise of success. It seems reasonable to place a far greater reliance on these newer models as our guiding conceptual lights and see what dark corners the lights begin to illuminate.

For far too long mental health systems, particularly mental hospitals, have been plagued by being treated as political footballs. Part of the reason for this uncomfortable state of affairs relates to the inability of professionals to precisely define and agree upon acceptable outcomes for various forms of treatment. Consequently, the resulting lack of demonstrable effectiveness has made us vulnerable to political maneuverings, both by evil men who want to keep certain types of deviancy under tight social control, and by honest men who are genuinely confused as to what mental health services should be. Most politicians are interested in promoting effective treatment systems. Yet, when professionals can't specify what it is they are doing, one shouldn't be surprised that these activities slip into a political realm and become highly-charged, emotional issues.

It seems obvious that the complexity of these problems and their ultimate solutions, demand a much stronger commitment to interdisciplinary efforts. A true interdisciplinary spirit probably has never been tried on a large scale in the area of mental health. We must find ways to include architects, engineers, political scientists, economists, and planners in the entire process of designing systems and delivering services.

Making predictions about the future, even though inherently reinforcing, is a very risky business; we need more projections about the kind of services likely to be needed by our society in the future. Books like Tofler's *Future Shock* have called our attention to the vast societal changes which are currently underway. Although we must begin to plan the types of services which will be needed in the next few decades, it is likely that mental health planners suffer from "future shock" just as much, and perhaps more, than other people.

Like it or not, the future is at hand. To meet it, we are equipped only with what we have and what we know. These entities are modifiable to the extent that we are willing to subject our knowledge and delivery systems to critical analysis, and to change based on that analysis.

It is our contention that behavior analysis and systems analysis are the most viable models presently available for building and critically analyzing new efforts in the human services. Without doubt each will be found to have its limits and liabilities, but as yet these are unknown. They are most likely to become known when these models are subjected to the complex and vigorous demands of mental health delivery systems.

The challenge is there. The quality of life for that increasing segment of our population who use human service organizations depends in large part upon our meeting this challenge.

Name Index

Subject Index